Einstein AND Buddha

THE PARALLEL SAYINGS

Einstein and Buddha is an inspired effort to meet the 21st-century challenge of developing a synthetic world view. McFarlane juxtaposes quotations from Eastern contemplatives and Western scientists with insight, clarity and intellectual integrity.

> —Dr. Ron Leonard, Department of Philosophy,
> University of Nevada–Las Vegas

Einstein and Buddha provides deep, simple and quotable insights that should help mend the rift between science and spirituality. If you put your thumbs over the quotation sources, you won't be able to tell who said what, when.

> —Fred Alan Wolf, Ph.D., physicist and author of
> *Mind into Matter, The Spiritual Universe*

This anthology provocatively illustrates the points of convergence between the quantitative investigation of the objects of consciousness and the qualitative exploration of consciousness itself.

> —B. Alan Wallace, author of
> *The Taboo of Subjectivity: Toward a New Science of Consciousness*

Many have explored the remarkable convergence between the mystical traditions of the world and modern science. However, none of them has done this in a more succinct and convincing way than *Einstein and Buddha*; this remarkable collection of quotes by famous Eastern mystics and modern physicists is a fascinating contribution to the emerging paradigm.

> —Stanislav Grof, M.D., author of
> *The Cosmic Game* and *Psychology of the Future*

P9-DTO-390

Einstein AND Buddha

THE PARALLEL SAYINGS

EDITOR

Thomas J. McFarlane

INTRODUCTION

Wes Nisker

Ulysses Press

First Ulysses Press Edition 2002

Published by:
ULYSSES PRESS
P.O. Box 3440
Berkeley, CA 94703
www.ulyssespress.com

The publishers have generously given permission to use quotations from the works cited on pages 175–77.

Library of Congress Cataloging-in-Publication Data

Einstein and Buddha : the parallel sayings / editor, Thomas J. McFarlane ; introduction, Wes Nisker.-- 1st Ulysses Press ed.

     p. cm.
   Includes bibliographical references.
   ISBN 1-56975-274-5 (alk. paper)
    1. Buddhism and science. 2. Buddhism--Quotations, maxims, etc. 3. Religion and science. 4. Scientists--Quotations. I. McFarlane, Thomas J.

BQ4570.S3 E36 2001
294.3'375--dc21
2001049727

ISBN10 1-56975-337-7 (paperback)
ISBN13 978-1-56975-337-8

Printed in Canada by Transcontinental Printing

10 9 8 7 6 5 4 3

Project Editor: Ray Riegert
Senior Editor: Richard Harris
Associate Editor: Steven Schwartz
Design: Leslie Henriques and Big Fish

Distributed by Publishers Group West

CONTENTS

Introduction	*vii*
Editor's Preface	*xi*
PARALLEL SAYINGS, PARALLEL WORLDS	3
THE PARALLEL SAYINGS	21
The Human Experience	23
The Touchstone of Truth	33
Paradox and Contradiction	43
Speakable and Unspeakable	53
Subject and Object	63
Name and Form	75
Illusions and Delusions	87
Waves, Fields and Energy	97
Particles and Matter	105
Wholeness and Interdependence	115
Time and Space	125
Manifestation and Causality	137
Unity and Plurality	147
Physics and Mysticism	157
Bibliography	*165*
About the Author	*176*

Acknowledgements

Among those who share credit for this book, foremost is Ray Riegert of Ulysses Press. It was Ray who conceived the original idea for *Einstein and Buddha*, and who entrusted me with the exciting challenge of searching through the vast scientific and spiritual literature to find and organize the parallel sayings. Credit also goes to the editorial team at Ulysses Press, and especially to Richard Harris. His significant contributions to the introductory sections of the book helped make the abstract ideas of physics and mysticism more accessible to the general reader. I would also like to acknowledge the editorial efforts of Steven Schwartz, who reviewed early drafts of the manuscript and provided valuable feedback.

Special thanks go to Mary Rosenbaum for handling the many permission letters required for a book such as this. I am also grateful to Marek Alboszta and Joel Morwood, who both read the entire manuscript and gave helpful feedback regarding its content and organization. In addition, Joel Morwood generously shared with me his vast database of quotations from the world's spiritual traditions, giving me a helpful jump-start on my extensive research. Finally, I give my warm thanks and appreciation to Agnieszka Alboszta, who gave me encouragement and moral support throughout the writing process, and reviewed the manuscript several times during its various stages of revision. Agnieszka made many suggestions to improve the clarity of the writing, and helped spot and correct many of my mistakes.

While credit goes to all these people for their kind contributions, any errors or shortcomings that still remain in this book belong to me.

—Thomas J. McFarlane

INTRODUCTION

By Wes Nisker

Einstein and Buddha: The Parallel Sayings is an exhilarating document—one that validates the wisdom of our species, unites cultures and ways of knowing, and suggests that humans might actually understand a little something about the nature of reality. What we also have here, in the words of both scientists and spiritual masters, are the outer limits of our understanding.

Einstein and Buddha represents an auspicious conjoining of two distinctly different ways of knowing. Generally speaking, from an overview of world culture it might seem as though the Earth was divided according to the two hemispheres of the brain. Asia was assigned the right hemisphere, and its great sages turned their attention inward, seeking truth through intuition and receptive quietude. In Europe and the Mediterranean—the left hemisphere—the search for truth turned outward, and became a process of deconstructing and analyzing the world, relying on the more aggressive powers of reason. The Asian wisdom traditions tended to see more wholistically, while the West was more interested in making distinctions. In our time, modern communications and travel have served as a corpus callosum, connecting the two hemispheres and revealing an astonishing agreement about the laws of nature and the structure of deep reality. Taken together, we now have what might be called "the full-brain approach."

Reading through this collection, I find myself especially thrilled to realize that modern physics has validated the insights of mystics and meditators. Even though the opposite is also true, our culture tends to be skeptical about personal revelation, viewing science as the highest authority, especially when it comes to describing how the cosmos works. It appears, however, if modern physics is to be taken seriously, that the great Asian spiritual masters were quite adept at seeing into matter, space-time, and the conundrum of consciousness.

The insights of the spiritual masters may seem quite surprising when you realize that they saw only with the naked mind, without the use of radio telescopes, atom smashing machines, or laser photography. Conversely, we should remember that it was that naked mind in the West that came to invent those powerful machines. Meanwhile, in the West we have long held a view of spiritual knowing as something that happens randomly, usually involving a lightning strike, a vision of a burning bush, or some such strange event. In recent decades, however, many Western scholars and seekers of truth have done extensive study in the Asian wisdom schools and have discovered that, like science, this other way of knowing involves a clearly proscribed and rigorous discipline. Although it may sound contradictory, we are realizing that mystical seeing can be learned.

A number of Western Buddhist teachers have even described the path of meditation as a form of "scientific" investigation, using very specific procedures that lead to recognizable, predictable results. In many schools of Buddhism, the meditator begins by developing the quality of "mindfulness," described as "a non-interfering, non-judgmental awareness," which is exactly the attitude a scientist is supposed to have when conducting an experiment. The meditator, like the scientist, attempts to become as objective as possible about what is being observed, even though in the meditator's case the subject is often one's own self. Furthermore, the meditative experiment is replicable with each new meditator who follows the procedures. As a meditation teacher, I can attest to the results: most meditators will arrive at similar insights into the nature of reality.

Although people will have different conceptual frameworks and therefore different ways to express what they see in meditation, their insights often involve descriptions of reality or the laws of nature. Through meditation, people can realize that "mind" is a co-creator of the world; that rather than any single cause of an event, all phenomena are connected in a web of relationship similar to what is described in "complexity theory"; that every perspective is relative to the observer; that energy comes in "quanta"; that all things are in process and there is no solidity anywhere. As you will see in *Einstein and Buddha*, these insights are in close agreement with the latest scientific theories.

There is perhaps one important difference between the truths of the scientist and those of the meditator or spiritual adept. While the scientist has discovered that everything is in constant flux by examining the external world, the meditator will discover this same truth inside his or her own mind and body, which makes the insight very personal. The truth of impermanence becomes relevant to the meditator's own life; knowledge about the workings of the universe turns into wisdom. Of course, the scientist's truth can be personally transformative as well, and nobody has proven this better than Einstein himself, but perhaps it happens more frequently in the Asian wisdom schools, where personal transformation is the whole point of the investigation in the first place.

The scientists and spiritual masters are just beginning to converse and compare notes, so any consequences or conclusions are premature. *Einstein and Buddha* is an excellent primer for their dialogue, and offers all of us a penetrating look at the basis for this meeting of the minds. One of the great challenges of our time is to unite reason with the human heart, cognition with compassion, science with spirituality, and here we have the groundwork. What better way to bring the twos together than through the words of the scientists and spiritual masters themselves? May these sayings lead to the benefit of all beings.

E D I T O R ' S P R E F A C E

"The concept of space detached from any physical content does not exist."
Or in other words:
"If there is only empty space, with no suns nor planets in it, then space loses its substantiality."

One of these statements was by Siddhartha Gautama, known as the Buddha, who lived around the 5th century BC and founded a major religion which today has 350 million followers worldwide. The other statement was written in the 20th century by Albert Einstein, whose theory of relativity revolutionized the science of physics. (To find out who said which statement, see page 136.)

Although the two men lived millennia apart, on opposite sides of the earth, and used different methods to understand and investigate the nature of reality, both discovered many of the same truths. Besides the nature of space, statements by Einstein and other modern physicists parallel those of Eastern mystics in such areas as the nature of time and matter, the wholeness of all things, the description of reality through paradox, the relationship between the observer and the observed and the need to test knowledge through experience. In many (though not all) aspects, modern physics leads us ever closer to the view of reality embraced by ancient Asian philosophies.

For the past 300 years, science and religion were viewed as antagonistic opposites. Parallels between Eastern religions and modern physics challenge this notion, and mark the first signs of a profound transformation in the way we investigate and understand reality. In future centuries, people may look back at our era as a second scientific revolution—one in which the duality of science versus religion was revealed to be an illusion, and humankind recognized the two disciplines as complementary rather than contradictory.

Today, we have taken our first steps down that road and do not know how the adventure will unfold. This book's purpose is to introduce the parallels between modern science and Eastern mysticism in a straightforward way that encourages the reader to examine these parallels and see what wonders they reveal.

Science and religion are neither entirely different nor entirely the same. Comparing the parallel sayings of physicists and mystics allows us to look beyond the antagonism that has traditionally existed between the two to a new era of harmony and integration.

These parallel sayings are provocative seeds for contemplation of a reality beyond the strictly physical or purely spiritual way of understanding. Where the sayings show similarities, they reveal the unity amid the differences between science and religion. Where they show differences, they illuminate the profound nature of that unity. Like Zen koans, they help loosen the mind's grasp on simplistic views of reality, opening to us a space where insight can dawn.

About the same time the Buddha lived in India, the first Western philosophers planted the seeds of science in ancient

Greece. Chief among them, Pythagoras proposed that the entire cosmos was mathematically ordered and could be rationally understood in terms of numbers and their relationships. Although this seed took centuries to sprout, it was—and still is—at the root of the scientific approach to understanding reality. Nearly 2000 years later, during the Renaissance, Pythagoras' notion was linked to the experimental method, and science as we know it was born.

Most historians would agree that animosity between science and religion in Western culture first arose when scientific thinkers like Copernicus and Galileo made discoveries that seemed to flatly contradict the religious dogma of the day. As a consequence, those who would further scientific knowledge were condemned as heretics by the church. Later, as the power of science increased, reality was divided into material and spiritual halves, and science was given authority over the material world. This artificial division between religion and science, in turn, gave rise to several misconceptions that are commonly held to this day.

One such misconception is that science deals with physical facts, while religion deals with spiritual values. Science, we are told, tells us what facts are true, while religion tells us what values are good. Yet when science makes a statement of fact, it is also making a judgment about what is valuable. If a golden nugget is found to be fool's gold, all its value suddenly vanishes; in the same way, if science declares the material world to be the only reality, it implicitly challenges the value of any spiritual reality. What we take to be scientific fact determines what we value. Also, what we value helps determine what we consider to be scientific facts. We see this especially when two very different theories explain the same experimental data. In that case, scientists choose one theory over the other by appeal to values such as beauty, elegance, simplicity and coherence.

Facts, as well as values, are also essential to religion. The essential core of any system of spiritual values is the true nature of reality. For mystics, Buddha nature, Brahman or the Tao are not just the ultimate value but the ultimate fact, reality and truth. Knowledge of this truth is the key to salvation or liberation. As Jesus said, "Ye shall know the truth, and the truth shall make you free." Or, as the Buddha put it, "One who acts on truth is happy in this world and beyond." Both facts and values are inseparable elements of religion as well as science.

Another misconception is that science deals with the outer, material world, while religion deals with the inner, spiritual world. Actually, scientific theories are shaped not only by our perceptions of the outer world but also by the scientist's aesthetic sensibilities such as mathematical beauty, elegance, simplicity and coherence. The ultimate object of scientific knowledge is not the external world seen through the senses but a theory of the world's order and harmony as perceived in the scientist's mind.

Although the mind is the primary tool used in religious contemplation, the scope of such contemplation includes both inner and outer objects. The ultimate goal of the spiritual seeker is to recognize that there is no real distinction between inner and outer, self and other. Religious traditions, as well as scientific thought, deal with both the internal world of insights and the external world of sensory phenomena.

Nor is it accurate to say that science is based on rational thought and doubt, while religion is based on intuition and faith. It is true that reason and logic are important standards of scientific thought; yet not a single great creative scientific breakthrough would have been possible without intuitive insight and inspiration. And while intuitive insight is the fountain of all sacred teachings, rational thought and philosophical reasoning play important roles in clearing the mind so intuition can shine without obstruction.

Similarly, although the scientist must be willing to doubt every physical hypothesis, the practice of science calls for fundamental faith that reality is rationally comprehensible. The spiritual seeker, too, must begin with faith that the ultimate truth can be known, but will never know that truth without a radical doubt of all human forms of knowledge. Both faith and doubt, as well as logic and intuition, are qualities that the scientist and the mystic share in common.

Artificial distinctions between science and religion, once dissolved, reveal the harmony between them. The Good and True are united, Plato tells us. The Upanishads declare that the inner Atman and the outer Brahman are identical. There is a deep level of reality that both outer and inner, fact and value, share in common. Knowing this reality requires both reason and intuition, doubt and faith. As two complementary approaches to the same reality, science and religion can combine their observational techniques with concepts and symbols that may give rise to revolutionary new cosmologies and theories.

Both scientists and mystics investigate reality by refining their capacities to observe extremely subtle phenomena far beyond the limits of ordinary perception. Physics constructs elaborate measuring devices and uses mathematical symbols to represent reality. The contemplative traditions cultivate special forms of insight through meditation and other mental disciplines and use myth, art, poetry, parable and philosophy to represent reality.

One difference is that mystics view their doctrines as ultimately pointing to a reality that cannot be known in concept or theory, while scientists are concerned with developing conceptual models of an objective reality. Their goal is not reality itself but a theory about reality. This difference, however, is not essential to an opposition between science and religion. Most spiritual teachings and experiences are not about the ineffable Absolute but relate to what can be objectively known. The same is true of science.

By comparing the scientific and contemplative approaches to reality, we can see the possibilities for convergence. Einstein and Buddha both sought to know the deepest truths about the same reality, using many of the same investigative principles. It's no wonder they had similar things to say about what they discovered.

Thomas McFarlane
November 2001

Parallel Sayings, Parallel Worlds

Skimming through this book, you will notice that it contains quotes by quite a few physicists besides Albert Einstein. The new philosophy underlying modern physics was not a mere eccentricity of this one wild-haired genius. The breakthroughs that would take science beyond the visible to the brink of the unimaginable were made also by other physicists—Einstein's contemporaries as well as a younger generation of physicists who developed quantum theory, observed subatomic particles and conceived the first nuclear weapons. Among them were such scientists as Niels Bohr, who incorporated the Yin-Yang symbol of Chinese cosmology in his coat of arms, and Robert Oppenheimer, who baffled his fellow scientists with quotes from the Bhagavad Gita.

As you'll see in the ensuing pages of this book, many of the things they said about the nature of reality were remarkably similar to the teachings of Gautama Buddha and to the writings of other Buddhists through the centuries, right up to people like the Dalai Lama, the Nobel Prize–winning supreme spiritual leader of Tibetan Buddhism, and D. T. Suzuki, who helped introduce Zen Buddhism to the United States in the mid-20th century.

Also represented here are mystics from other Eastern religious traditions, particularly Hinduism and Taoism, which place the

Buddhist phenomenon in context. When the Buddha set forth on his spiritual journey, he lived for six years among Hindu ascetics. The framework for his spiritual discoveries derived from this experience in much the same way that Jesus' message was grounded in his devout Jewish background. Buddha's message spread far beyond the borders of India, reaching across Tibet, China, Southeast Asia and Japan and even today continuing its growth into the Western Hemisphere and Europe. Like Christianity, it often grew by assimilating regional religious beliefs and practices. In China, the complementary philosophies of Taoism and Confucianism held sway and helped shape the evolving Buddhist philosophy.

THE PHYSICISTS

Among the cast of characters who appear in this book, the lead actor is undoubtedly Albert Einstein. In 1905, at age 26, he rocked the scientific world with his special theory of relativity, showing time and space to be dependent on the observer. His famous equation $E=mc^2$ revealed matter and energy as interchangeable forms of the same substance. Ten years later he devised the general theory of relativity, explaining gravity not as a force but rather as the warping of space-time. Before Einstein, physicists had viewed time and space as completely separate from each other, matter and energy as fundamentally different, and gravitation as a mysterious force that acted at a distance through empty space. Einstein's work proved all these beliefs wrong, revolutionizing basic concepts of reality.

Like many other pioneers of modern physics, Einstein taught in German universities until the Nazis came to power. In 1933, he fled to the United States and spent the rest of his active life at Princeton University's Institute for Advanced Study in New Jersey. There, he continued the search for a unified field theory that would unite space, time and gravity with electromagnetism; more years of his professional life were spent on this than any other project.

Throughout his life, Einstein was concerned with philosophical as well as scientific questions. He was deeply concerned with the human condition, social injustices, and virtues such as selflessness and devotion to higher ideals. His religious views were largely influenced by the 17th-century philosopher Baruch Spinoza, whose life paralleled Einstein's in many respects. Like Einstein's theories, Spinoza's greatest work, *Ethics*, expressed its rational mysticism behind logical abstractions while concluding that nature's harmony proves God's existence, and that elegant simplicity—the "oneness" Einstein had long hoped to find in the elusive unified field theory—holds the key to all understanding. As Einstein wrote, "All religions, arts and sciences are branches of the same tree. All these aspirations are directed toward ennobling man's life, lifting it from the sphere of mere physical existence and leading the individual toward freedom."

At the same time as Einstein was doing his early work, other physicists in Germany and other parts of Europe were beginning to probe the microscopic world of the atom. As they did, they found that the classical physical laws of mechanics, electrodynamics and thermodynamics no longer fit the experimental data. They saw particles of matter behave like waves and waves of light behave like particles. Strict laws of cause and effect gave way to such spontaneous, unpredictable events as radioactive decay. And atoms, they discovered, only absorbed and emitted energy in discrete chunks, called quanta. This last feature gave its name to the system that finally explained it all: quantum theory.

Max Planck, a German physicist, had produced the earliest hints of quantum physics with his radiation formula, which first introduced the quantum principle into physics. A complete theoretical deduction of the formula showed classical notions of the atom to be flawed and laid the foundation for the work of Bohr and his colleagues in other parts of Europe and the United States.

Among them, the Danish physicist Niels Bohr played a central role in both the discovery and development of quantum theory. His correspondence principle showed the way to relate new quantum laws coherently to established laws of classical physics. Bohr's principle of complementarity allowed physicists to make sense of paradoxes such as the fact that subatomic particles had both wave and particle properties. Second only to Einstein among 20th-century physicists, the Danish scientist participated in a long-running scientific debate with Einstein over the significance of quantum theory. Einstein refused to accept that nature is fundamentally random and devised "thought experiments" to support his maxim that "God does not play dice with the universe," as Bohr's interpretation of quantum theory seemed to imply. In each instance, Bohr showed Einstein where he had gone wrong. He also likened modern physics to Eastern mysticism in these words: "For a parallel to the lesson of atomic theory, [we must turn] to those kinds of problems with which thinkers like the Buddha and Lao Tzu have been confronted, when trying to harmonize our position as spectators and actors in the great drama of existence."

Werner Heisenberg, one of Bohr's students, discovered the mathematical laws of quantum theory in 1925, and formulated the famous Heisenberg uncertainty principle in 1927. According to this principle, it is impossible to measure both the exact position and velocity of a particle at the same time. During World War II, Heisenberg remained in Germany as the head of Hitler's unsuccessful nuclear weapons program. He was interned in England after the war but eventually returned to Germany to continue his theoretical research as director of the Max Planck Institute. Heisenberg's philosophical views can best be described as a modern scientific form of Plato's rational mysticism. "The search for the 'one,' for the ultimate source of all understanding," he wrote, "has doubtless played a similar role in the origin of both religion and science."

Wolfgang Pauli, a young prodigy who worked as Bohr's assistant, is best known for the Pauli exclusion principle. Developed in 1925, the principle held that no two electrons in an atom can have the same attributes. Although a brilliant scientific success, Pauli's personal life was troubled by his mother's suicide, a bitter divorce and a drinking problem. He sought help from Swiss psychoanalyst Carl Jung, who shared his own mystical views with Pauli throughout many years of correspondence. Pauli once wrote, "I still regard the conceptual aim of overcoming the contrasts, an aim which includes a synthesis embracing the rational understanding as well as the mystic experience of one-ness, as the expressed or unspoken mythos of our own present age."

Erwin Schrödinger, an Austrian who was originally more interested in a career as a philosopher than a physicist, developed a different formulation of quantum theory around the same time as Heisenberg. His wave equation described electrons not as individual, localized particles of matter orbiting an atom's nucleus like planets around the sun but rather as standing waves of probability—mere ghosts of particles in the classical sense. Only when such a particle is observed can it be said that it exists in a particular position; unobserved, its position has no existence, only a potentiality for existence.

As Einstein, Bohr and their colleagues were doing their most revolutionary work in continental Europe, British scientists were making their own important contributions. Foremost among them was Sir Arthur Eddington, director of the Cambridge Observatory. Mounting an expedition to Africa to photograph a total solar eclipse in 1919, he produced the first confirmation of Einstein's theory that gravity bends light. A Quaker by upbringing, Eddington was intrigued with the philosophical implications of relativity and quantum theory and authored several books on the subject. In one, Eddington observed that "As truly as the mystic, the scientist is following a light; and it is not a false or an inferior light."

Sir James Jeans, one of Eddington's leading British colleagues, also wrote about physics and philosophy. Jeans is remembered for coining the term "new physics" to describe the work of Bohr, Heisenberg, Schrödinger and others. He did not include Einstein among the "new physicists" but saw him at the end of the "mechanical age" that had started with Sir Isaac Newton.

The younger generation of physicists, such as Heisenberg and Pauli, were raised between the World Wars, and faced the moral dilemma of applying physics to inventing weapons of mass destruction.

Robert Oppenheimer, a leading American physicist educated in Europe, was appointed by President Franklin Delano Roosevelt to serve as director of the Manhattan Project, the effort to build the first atomic bomb, at a top-secret installation in Los Alamos, New Mexico. After the war he chaired the Atomic Energy Commission Advisory Committee, where he opposed building larger, more powerful hydrogen bombs—a politically unpopular stance in the early years of the Cold War. In 1953 he was accused of having been a communist sympathizer. Although cleared of treason charges, his security clearance was taken away, along with his government position. He spent his later years as director of Princeton's prestigious Institute for Advanced Study where Einstein had worked for 20 years. Looking back on his distinguished career in physics and its relation to the world, Oppenheimer mused, "The general notions about human understanding illustrated by discoveries in atomic physics are not in the nature of things totally unfamiliar, wholly unheard of, nor new. Even in our own culture they have a history, and in Buddhist and Hindu thought a more considerable and central place. What we shall find is an exemplification, an encouragement, and a refinement of old wisdom."

Richard Feynman, one of the most brilliant and eccentric of American physicists, was sent to work on the Manhattan Project at age 24. After the war, he made important contributions to modern physics and won the Nobel Prize for his work. He is also celebrated as one of the best teachers of physics, and is remembered for leading the investigation of the Challenger space shuttle disaster. His books included the bestsellers *Surely You're Joking, Mr. Feynman* and *What Do You Care What Other People Think?*

David Bohm, perhaps the most unusual of the midcentury physicists, worked with the foundations of physics and quantum theory, and authored a respected textbook on quantum theory. Forced to leave the United States during the McCarthy era of the 1950s because he refused as a matter of principle to testify in the congressional hearings against his colleagues such as Oppenheimer, he became an expatriate in Great Britain. In the 1970s, Bohm initiated a long collaboration and dialogue with the Indian mystic Krishnamurti, and later became a friend of the 14th Dalai Lama, who referred to Bohm as "my physics teacher."

In the last decades of the 20th century, some physicists began writing explicitly about Eastern mysticism as a key to reconciling the paradoxes of the New Physics, and in so doing, pointed the way to the next step in the evolution of human consciousness.

First and foremost among them, Fritjof Capra earned his doctorate in particle physics from the University of Vienna in 1966 but soon turned his rational mind toward the challenge of comprehending an essentially mystical experience, which he describes in the preface to his now-classic 1976 book, *The Tao of Physics: An Exploration of the Parallels between Modern Physics and Eastern Mysticism.* Sitting by the ocean one afternoon, watching the waves wash against the shore, the California-based physicist realized that the vibrating molecules and atoms composing the scene around him were part of a cosmic dance of energy. "I felt its rhythm and I 'heard' its sound," he recalls, "and at that moment I knew that this was the Dance of Shiva, the Lord of Dancers worshipped by the Hindus."

Other writers soon followed Capra's intellectual-spiritual lead with more books exploring the parallels between the new physics and ancient mysticism. In 1986, philosopher Renee Weber wrote *Dialogues with Scientists and Sages,* in which she presented her conversations with such diverse thinkers as the Dalai Lama, J. Krishnamurti, David Bohm and Stephen Hawking. In 1995, Victor Mansfield, professor of physics and astronomy at Colgate University merged his scientific research with his studies in Tibetan Buddhism and Jungian psychology in *Synchronicity, Science and Soul-Making.* Amit Goswami, professor of physics at the University of Oregon, described how Eastern mysticism resolves the paradoxes of quantum theory in his book *The Self-Aware Universe.*

Another author who has given voice to the connection between physics and spirituality is Gary Zukav, a non-scientist who describes himself as a "Harvard grad, Vietnam vet, Green Beret, recovering soul and favorite Oprah guest." His book *The Dancing Wu Li Masters* (1984) stands as a classic in the field. In his foreword to a later work, *The Seat of the Soul*, Zukav describes how Niels Bohr and Albert Einstein "saw more than they could express through the language of...physics, and they sought to share what they saw.... They were mystics."

The book you are holding now, *Einstein and Buddha: The Parallel Sayings*, presents samples of the original sources that inspired Capra, Weber, Mansfield, Zukav and others, in the hope that it will let you share some of the sense of wonder they have brought to the new physics.

THE MYSTICS

One of history's most influential religious thinkers, Siddhartha Gautama, now called the Buddha (meaning "Awakened One"), grew up in luxury as the son of the King of the Shakya clan in northern India in the 5th century BC. In young adulthood he came to a spiritual crisis because of the human suffering he saw around him and left home to live and study with Hindu ascetics. After six years, it is said, he attained nirvana, the rare state of beatitude that transcends suffering and karma. Thereafter, his teachings diverged from mainstream Hindu traditions of the time, and from age 35 until his death at age 80 he wandered northern India as a spiritual teacher of the principles that would form the foundation of Buddhism. Although his following during his lifetime was limited to two provinces in India, his teachings spread throughout eastern Asia. Today Buddhism is one of the world's largest religions, with many more followers outside India than within.

At the core of Buddha's teaching are the Four Noble Truths:
- Life as we know it normally brings dissatisfaction and suffering.
- This suffering has a cause, so it is not an intrinsic or necessary part of the nature of things.
- The cause of suffering is our habitual, unconscious clinging to impermanent things as if they had some permanent essence.
- There is a path to end this habitual clinging to a mistaken view of things and so be free from the suffering that results from it.

Although various Buddhist teachings and doctrines developed as it spread into new regions—Theravada Buddhism in Sri Lanka and Southeast Asia, and the various forms of Mahayana Buddhism in Tibet, China, Korea and Japan—the Buddha's original teachings remain common to them all.

Over the millennia since, many other teachers, mystics and monks have contributed to Buddhist philosophy. The process enjoys a special continuity because the belief in reincarnation allows for Buddhism's greatest keepers of wisdom to remain among their followers, in new physical bodies, across the generations.

Among the other spiritual leaders quoted in this book is Ashvaghosha, who lived in central India during the 2nd century AD. He was originally a Brahmin, a member of the highest Hindu caste from which priests and intellectuals came. After conversion to Buddhism, he became instrumental in spreading the faith throughout India.

Siddha Nagarjuna also lived in the 2nd or 3rd century AD. A legendary figure, he is said to have displayed astonishing magical powers during childhood. Nagarjuna coined the concept of "the Middle Way," which is central to Buddhist thought, and systematically developed the important concept of "emptiness." He is considered the greatest Buddhist philosopher ever.

In the 19th century, one of the foremost Buddhist philosophers and teachers of his day was the Tibetan monk Kongtrul Lodrö Tayé. Also among modern authorities on Buddhism is Gen Lamrimpa, who has written extensively on Buddhist meditation practices. Then there are K. Venkata Ramanan, a modern expositor of Nagarjuna's teachings, and Daisetz Teitaro (D.T.) Suzuki, who is credited with introducing Zen Buddhism to America in the mid-20th century.

Certainly the most familiar Buddhist figure in modern times is Tenzin Gyatso, the 14th Dalai Lama, spiritual leader of Tibetan Buddhism. While still in his 20s he went into exile in northern India after the Maoist Chinese invasion of Tibet. Today the Dalai Lama resides in India and travels the world as a teacher and best-selling author, conducting discussions with everyone from Hopi

Indians to American presidents. He was awarded the Nobel Peace Prize in 1989.

Hinduism is the most ancient and diverse of the major Eastern religious traditions. Like Buddhism, the core of the Hindu traditions is the essential mystical teaching that reality is a harmonious unity with no real divisions, and diversity is only the impermanent, cyclical play of this fundamental truth. Buddha began his path to enlightenment as an ascetic following the Hindu tradition. Other important figures over the centuries also started their careers as Hindus, often of the priestly Brahmin caste, and later converted to Buddhism, and the history of each religion has been shaped to a great extent by the other over the centuries.

While the sacred writings of Hinduism, like those of Buddhism, literally fill libraries, a few ancient works are central to Hindu thought. Thousands of years ago in ancient India, Vedic seers transmitted their spiritual knowledge through the sacred hymns of the Rig Veda. Over the centuries, these hymns were supplemented by more philosophical teachings, notably the Upanishads. These 24 texts contained, among other things, instructions for spiritual practice and descriptions of the true nature of reality. Another of the most important Hindu religious texts, the Bhagavad Gita ("Lord's Song") is Book Six of the Mahabarata, Hinduism's "Old Testament." This multi-volume epic was written over a span of time from the 5th century BC to the 2nd century AD. Collecting myths and legends into a unified history going all the way back to creation, the Mahabarata is said to be the longest poem ever written and one of the greatest masterpieces of world literature.

The sacred scriptures of Hinduism do not describe a fixed doctrine. Instead, it is an inherently dynamic and living religion whose doctrines are constantly renewed, extended and interpreted by the mystics of each generation.

Shankara, one of the most influential mystics of classical Hinduism, lived during the 9th century AD, a time when Buddhism had grown to be India's major religion and ancient Hindu practices had become fragmented. Shankara traveled throughout the country, reunifying the Hindu faith under a systematic, nondual philosophy based on the Upanishads and called Advaita Vedanta. According to Shankara, all dualism is illusory and not real. The only reality is Brahman, the "One without a second." The innermost essence of every person, the Atman, is ultimately identical with the Brahman. To know the Atman at the core of oneself is to know Brahman as well.

Abhinavagupta, a philosopher and mystic in the Kashmir Shaivism stream of the Hindu tradition, continued Shankara's effort to unify Hinduism as a living, growing alternative to Buddhism. A great writer, he is credited with reviving the Sanskrit literary tradition that began in ancient times with the Mahabarata.

Hinduism continues as a vibrant religion today. One recent mystic, Ramana Maharshi, came to a realization of his deepest Self as a child. After spending many years in silence, he emerged to teach the Advaita Vedanta doctrine. He is revered as one of the greatest saints of modern times, and a growing number of spiritual teachers trace their lineage back to him.

Sri Aurobindo, another influential Hindu mystic, started his career as an academic and public servant in colonial India and became a political leader. Imprisoned for sedition, he began to practice yoga, and after his release he went into self-imposed seclusion to continue his spiritual work. His integral approach to spiritual practice synthesized the main teachings of Hinduism into a single system. He reported his metaphysical insights in a magazine he published and later in books. His unique philosophy of spiritual evolution has influenced contemporary western philosophers.

In China, about the same time the Buddha lived in India, the sage Lao Tzu was composing the seminal teachings of Taoism in his masterwork, the Tao Te Ching. In contrast to the principles of his contemporary, Confucius, who was more concerned with morality, society and politics, Lao Tzu focused on the personal, mystical and metaphysical aspects of life. Although it might be interpreted literally as a handbook for government, the Tao Te Ching can also be seen as an allegory of the soul, in a similar vein to Plato's Republic, describing the way of governing one's own soul and bringing it into harmony with the Tao.

Chuang Tzu, who lived in the 4th century BC, is considered second only to Lao Tzu in his influence on Taoism. He wrote discourses and parables explicitly presenting a personal philosophy for the individual. He taught that observing nature at work and reconciling polar opposites point the way to the Tao, where all dualities are resolved into unity. Chuang Tzu is called the "butterfly philosopher" because he wrote that he once dreamed he was a butterfly. When he awoke, he no longer knew if he was a man who had dreamed he was a butterfly—or a butterfly dreaming he was a man.

Taoist teachings are rich with paradox, describing the sage as humble, yielding, spontaneous and free. Without taking any action, the sage leaves nothing undone. All his power comes from surrendering his will to the Tao, attaining harmony with the natural order of all things. Taoism emphasizes the interdependence of opposites, which forever transform into each other according to the dynamic cycles of nature. The Tao is the mysterious, ineffable origin of Heaven and Earth, Being and Nothingness. Like Hindu liberation and Buddhist enlightenment, the Tao is identified with the fundamental reality underlying all change.

⁓ ⁓

These mystics and physicists are the main thinkers from two areas that have always seemed worlds apart. Removed thousand of years from each other in time and thousands of miles in space, they came to symbolize the great split in world history—East and West—and the great split in world thinking—Science and Religion. But in the following parallel sayings we will see just how close together they really were. Dealing with questions of time and space, cause and effect, paradox and contradiction, subject and object, name and form, wholeness and interdependence, the nature of matter and other topics, they used different words to say exactly the same thing.

The Parallel Sayings

The Human Experience

We are sometimes tempted to think of Einstein and other physicists as coldly analytical men because they expressed their thoughts through mathematics—a universal and precise language that is completely lacking in passion.

Yet when we peek behind the equations to look at these scientists' personal experiences and insights, we find that they were deeply moved by the music of the cosmos. In much the same way as mystics in the Asian tradition, great physicists like Einstein have dedicated themselves selflessly to higher ideas, relying upon intuition, beauty, simplicity and harmony as spiritual guides in their quest for truth. When they succeeded in glimpsing the way the universe works, they felt the same sort of amazement and exhilaration that followers of the Buddha experience in deep meditation.

For the scientist, as for the mystic, beauty goes hand-in-hand with truth. The harmony of diverse elements transforms worldly dissonance into a unity that the soul recognizes like an old friend.

Human beings can attain a worthy and harmonious life only if they are able to rid themselves, within the limits of human nature, of striving to fulfill wishes of the material kind.

ALBERT EINSTEIN

Men, driven on by thirst, run about like a snared hare; let therefore the mendicant drive out thirst, by striving after passionlessness for himself.

BUDDHA

The true value of a human being is determined primarily by the measure and the sense in which he has attained to liberation from the self.

ALBERT EINSTEIN

The Perfect Man has no self.

CHUANG TZU

A human being...experiences himself, his thoughts and feelings as something separated from the rest—a kind of optical illusion of his consciousness. This delusion is a kind of prison for us, restricting us to our personal desires and to affection for a few persons nearest to us. Our task must be to free ourselves from this prison by widening our circle of understanding and compassion to embrace all living creatures and the whole of nature in its beauty.

ALBERT EINSTEIN

True happiness comes not from a limited concern for one's own well-being, or that of those one feels close to, but from developing love and compassion for all sentient beings.

THE DALAI LAMA

The high destiny of the individual is to serve rather
than to rule.

ALBERT EINSTEIN

By merely doing actions in my service
Thou shalt attain perfection.

THE BHAGAVAD GITA

It is the very essence of our striving for understanding that, on the one hand, it attempts to encompass the great and complex variety of man's experience, and that on the other, it looks for simplicity and economy in the basic assumptions. The belief that these two objectives can exist side by side is, in view of the primitive state of our scientific knowledge, a matter of faith.

ALBERT EINSTEIN

He takes part in ten thousand ages and achieves simplicity in oneness. For him, all the ten thousand things are what they are, and thus they enfold each other.

CHUANG TZU

In our knowledge of physical nature we have penetrated so far that...here is neither suffering nor evil nor deficiency, but perfection only.

HERMANN WEYL

Everything is essentially consciousness, purity and joy.

SHANKARA

A mathematical truth is timeless, it does not come into being when we discover it. Yet its discovery is a very real event.

ERWIN SCHRÖDINGER

Realization is nothing to be gained afresh; it is already there. All that is necessary is to get rid of the thought 'I have not realized'.

SRI RAMANA MAHARSHI

Failure and deprivation are the best educators and
purifiers.

ALBERT EINSTEIN

Difficult circumstances...appear to be extremely
unfavorable to the practice of spiritual development.
However, for those transforming their outlook, especially
by cultivating the awakening mind, these situations become
an encouragement for the accomplishment of the practice.

GESHE RABTEN

Those who are not shocked when they first come across quantum theory cannot possibly have understood it.

NIELS BOHR

Is it not shocking to know that...all the heavens including all the luminaries whose lights are measured to reach this earth after millions of years are said to be mere bubbles in the ocean of eternal Emptiness?

D. T. SUZUKI

The Touchstone of Truth

Experience is the foundation of science. Every theory must be experimentally tested, and if it fails, it must be discarded or modified. Scientists not only devise experiments to verify their own theories but also to expose any flaws. The constant challenge of experience saves science from becoming a fixed dogma or a maze of unfounded speculations.

The mystic, too, speaks with the authority of experience. Like the scientist, the mystic seeks to clear away hidden biases, assumptions and illusions so that he or she can view subtle aspects of reality that are normally hidden.

The experiences of the mystic, like those of the scientist, reach beyond the ordinary. Testing religious beliefs through direct experience rather than accepting them on faith, the mystic relies on ways of knowing that may appear far different from those of science. Yet through direct investigation, the mystic and the scientist alike find reality's deeper truths.

Truth is what stands the test of experience.

ALBERT EINSTEIN

The real meaning of the Dharma...must be directly experienced.

SIDDHA NAGARJUNA

Let us get down to bedrock facts. The beginning of every act of knowing, and therefore the starting-point of every science, must be in our own personal experience.

MAX PLANCK

Personal experience is...the foundation of Buddhist philosophy. In this sense Buddhism is radical empiricism or experimentalism.

D. T. SUZUKI

The common root from which scientific and all other knowledge must arise...is the content of my consciousness.

SIR ARTHUR EDDINGTON

The Truth itself...can only be self-realized within one's own deepest consciousness.

BUDDHA

Science...is based on personal experience, or on the experience of others, reliably reported.

WERNER HEISENBERG

From the lips of your teacher you have learned of the truth of Brahman as it is revealed in the scriptures. Now you must realize that truth directly and immediately. Then only will your heart be free from any doubt.

SHANKARA

Experimenters search most diligently, and with the greatest effort, in exactly those places where it seems most likely that we can prove our theories wrong. In other words we are trying to prove ourselves wrong as quickly as possible, because only in that way can we find progress.

RICHARD P. FEYNMAN

In our world error is continually the handmaid and pathfinder of Truth; for error is really a half-truth that stumbles because of its limitations; often it is Truth that wears a disguise in order to arrive unobserved near to its goal.

SRI AUROBINDO

It is difficult for the matter-of-fact physicist to accept the view that the substratum of everything is of mental character. But no one can deny that mind is the first and most direct thing in our experience, and all else is remote inference.

SIR ARTHUR EDDINGTON

The external world is only a manifestation of the activities of the mind itself, and...the mind grasps it as an external world simply because of its habit of discrimination and false-reasoning. The disciple must get into the habit of looking at things truthfully.

BUDDHA

We should not let everything else atrophy in favor of the one organ of rational analysis.... It is a matter, rather, of seizing upon reality with all the organs that are given to us, and trusting that this reality will then also reflect the essence of things, the "one, the good and the true."

WERNER HEISENBERG

Transcendental intelligence rises when the intellectual mind reaches its limit and if things are to be realized in their true and essential nature, its processes of thinking must be transcended by an appeal to some higher faculty of cognition.

BUDDHA

We can only reason from data and the ultimate data must be given to us by a non-reasoning process—a self-knowledge of that which is in our consciousness.

SIR ARTHUR EDDINGTON

⸻

Emptiness is the result of an intuition and not the outcome of reasoning.... It is the Praja that sees into all the implications of Emptiness, and not the intellect.

D. T. SUZUKI

Paradox and Contradiction

Worldly conceptions of ordinary objects and everyday experience suddenly broke down when physicists began to explore the nature of the atom. The quantum world, where distinctions like "waves" versus "particles" blur, compels science to talk in terms of paradoxes.

In Einstein's world of relativity, the word "simultaneous" has no meaning. Time has no absolute standard. Twins can take separate journeys, in theory, and ultimately meet to find that one of them is an old man while the other is a child. In coming to terms with such a strange world, physicists have learned to accept paradox and view opposing concepts as complementary.

Eastern mystics would hardly be surprised. Their explorations beyond ordinary experience have long revealed that human ideas of the world are merely imperfect tools for viewing subtle phenomena. Far from being vague and inconsistent, paradox compels the mind to look beyond common conceptions and see the true nature of things.

If we ask, for instance, whether the position of the electron remains the same, we must say "no"; if we ask whether the position of the electron changes with time, we must say "no"; if we ask whether the electron is at rest, we must say "no"; if we ask whether it is in motion, we must say "no."

J. ROBERT OPPENHEIMER

He is far and he is near,
He moves and he moves not.

THE BHAGAVAD GITA

The classical concepts, i.e., "wave" and "corpuscle"...do not fully describe the real world and are, moreover, complementary in part, and hence contradictory.... Nor can we avoid occasional contradictions; nevertheless, the images help us to draw nearer to the real facts. Their existence no one should deny. "Truth dwells in the deeps."

NIELS BOHR

But if a man becomes attached to the literal meaning of words and holds fast to the illusion that words and meaning are in agreement, especially in such things as Nirvana which is unborn and undying...then he will fail to understand the true meaning and will become entangled in assertions and refutations.

BUDDHA

It is a primitive form of thought that things either exist or do not exist.

SIR ARTHUR EDDINGTON

To say "it is" is to grasp for permanence. To say "it is not" is to adopt the view of nihilism. Therefore a wise person does not say "exists" or "does not exist."

SIDDHA NAGARJUNA

If I say they behave like particles I give the wrong impression; also if I say they behave like waves. They behave in their own inimitable way, which technically could be called a quantum mechanical way. They behave in a way that is like nothing that you have ever seen before.

RICHARD P. FEYNMAN

Now do you say that you are going to make Right your master and do away with Wrong, or make Order your master and do away with Disorder? If you do, then you have not understood.... Obviously it is impossible.

CHUANG TZU

The idea of complementarity in modern physics has demonstrated to us, in a new kind of synthesis, that the contradiction in the applications of old contrasting conceptions (such as particle and wave) is only apparent.... The only acceptable point of view appears to be the one that recognizes both sides of reality—the quantitative and the qualitative, the physical and the psychical—as compatible with each other, and can embrace them simultaneously.

WOLFGANG PAULI

The world for those who have gained a satori is no more the old world as it used to be.... Logically stated, all its opposites and contradictions are united and harmonized into a consistent organic whole.

D. T. SUZUKI

Everything—anything at all—is at the same time particle and field.

ERWIN SCHRÖDINGER

Everything has its "that," everything has its "this." A state in which "this" and "that" no longer find their opposites is called the hinge of the Way.

CHUANG TZU

Bohr advocated the use of both pictures, which he called complementary to each other. The two pictures are of course mutually exclusive, because a certain thing cannot at the same time be a particle (i.e., substance confined to a very small volume) and a wave (i.e., a field spread out over a large space), but the two complement each other.

WERNER HEISENBERG

False-imagination teaches that such things as light and shade, long and short, black and white are different and are to be discriminated; but they are not independent of each other; they are only different aspects of the same thing, they are terms of relation not of reality.

BUDDHA

Yet all those who have truly understood quantum theory...look upon it as a unified description of atomic phenomena, even though it has to wear different faces.

NIELS BOHR

For we see that these apparently opposite terms of One and Many, Form and the Formless, Finite and Infinite, are not so much opposites as complements of each other; not alternating values...but double and concurrent values which explain each other; not hopelessly incompatible alternatives, but two faces of the one Reality.

SRI AUROBINDO

Speakable and Unspeakable

In modern physics, the more we seem to know about reality, the more reality seems to transcend our abilities to know it. Quantum physicists must confess to knowing much less about the position of an electron than classical physicists a century ago thought they knew. Scientific breakthroughs often compel scientists to recognize that reality transcends theory, reason and observation. Even as they strive to build more accurate and comprehensive models of reality, physicists find that, more and more, reality eludes them. It becomes essential to move beyond language and thought.

As Eastern mystics attest, ultimate knowledge of reality lies infinitely beyond the reach of the thinking mind and its philosophical systems. Such useful stepping stones can become barriers to knowledge if we forget that they are pointers to the truth, not truth itself. The Buddha likened his teachings to a raft his disciples could use to cross a river. Once the river was crossed, the raft would be of no further use and should be left behind.

We must be clear that, when it comes to atoms, language can be used only as in poetry.

NIELS BOHR

Truth cannot be cut up into pieces and arranged in a system. The words can only be used as a figure of speech.

BUDDHA

One may say that the human ability to understand may be in a certain sense unlimited. But the existing scientific concepts cover always only a very limited part of reality, and the other part that has not yet been understood is infinite.

WERNER HEISENBERG

Your life has a limit but knowledge has none. If you use what is limited to pursue what has no limit, you will be in danger.

CHUANG TZU

All our science, measured against reality, is primitive and childlike.

ALBERT EINSTEIN

Calculate what man knows and it cannot compare to what he does not know.

CHUANG TZU

This world faces us with the impossibility of knowing it directly.... It is a world whose nature cannot be comprehended by our human powers of mental conception.

MAX PLANCK

Brahman is...outside the range of any mental conception.

SHANKARA

There can be no descriptive account of the structure of the atom; all such accounts must necessarily be based on classical concepts which, as we saw, no longer apply.

NIELS BOHR

There the eye goes not, nor words, nor mind. We know not, we cannot understand, how he can be explained.

THE UPANISHADS

The experimental physicist must be able to talk about his experiments and therefore he is forced to employ the concepts of classical physics, although he realizes full well that they provide an inadequate description of nature. This is his fundamental dilemma, and one he cannot simply dismiss.

WOLFGANG PAULI

The contradiction so puzzling to the ordinary way of thinking comes from the fact that we have to use language to communicate our inner experience which in its very nature transcends linguistics.

D. T. SUZUKI

In quantum theory, we are beyond the reach of pictorial visualization.

NIELS BOHR

Self-realization is an exalted state of inner attainment which transcends all...illustrations.

BUDDHA

Thought as we know it would be impossible unless we could express its results in logical terms. Yet, the basic thinking process probably cannot be described as logical.

DAVID BOHM

Ordinary logic is the most useful implement in our practical life...[but] the spirit or that which occupies the deepest part of our being requires something thoroughly non-conceptual, i.e., something immediate and far more penetrating than mere intellection.

D. T. SUZUKI

The belief in an external world independent of the perceiving subject is the basis of all natural science. Since, however, sense perception only gives information of this external world or of "physical reality" indirectly, we can only grasp the latter by speculative means. It follows from this that our notions of physical reality can never be final. We must always be ready to change these notions—that is to say, the axiomatic basis of physics—in order to do justice to perceived facts in the most perfect way logically.

ALBERT EINSTEIN

While the Tathagata, in his teaching, constantly makes use of conceptions and ideas about them, disciples should keep in mind the unreality of all such conceptions and ideas. They should recall that the Tathagata, in making use of them in explaining the Dharma always uses them in the semblance of a raft that is of use only to cross a river. As the raft is of no further use after the river is crossed, it should be discarded. So these arbitrary conceptions of things and about things should be wholly given up as one attains enlightenment.

BUDDHA

Subject and Object

In the world of modern physics, we do not see an objective reality in itself. Instead, we see a reality created in the very act of observation by the meeting of subject and object. We participate in bringing about the properties of the objects we observe. According to Einstein's Special Theory of Relativity, an object's properties such as mass and length have definite values only when an observer is specified. Classical physicists of the 19th century would argue that these values always exist, even though they have not been measured.

The imaginary division between subject and object has long been a fundamental principle of Eastern contemplative traditions. The Buddha spoke of the emptiness of both subject and object—the self and the world. The Hindu Upanishads similarly identify Atman, the ultimate nature of the self, with Brahman, the ultimate nature of the world. Neither exists independent of the other, and both are inseparable aspects of a single, nondual reality.

Every man's world picture is and always remains a
construct of his mind and cannot be proved to have any
other existence.

ERWIN SCHRÖDINGER

The objective world rises from the mind itself.

BUDDHA

The physical world is entirely abstract and without "actuality" apart from its linkage to consciousness.

SIR ARTHUR EDDINGTON

Separated from the mind there are not objects of senses.

ASHVAGHOSHA

The common division of the world into subject and object, inner world and outer world, body and soul, is no longer adequate and leads us into difficulties.

WERNER HEISENBERG

The once believed ultimacy of the line of division between the "self" and the "not-self," the subjective and the objective, is rejected as untrue.

K. VENKATA RAMANAN

An independent reality in the ordinary physical sense can neither be ascribed to the phenomena nor to the agencies of observation.

NIELS BOHR

Being under illusion means perceiving objective appearances and mental appearances as having independent reality.

BOKAR RINPOCHE

Everything material is also mental and everything mental is also material.

DAVID BOHM

Things are objects because of the subject [mind]; the mind [subject] is such because of things [object].

SENGTSAN

Subject and object are only one. The barrier between them cannot be said to have broken down as a result of recent experience in the physical sciences, for this barrier does not exist.

ERWIN SCHRÖDINGER

Subjectivity and objectivity are only two sides of one consciousness.

SRI AUROBINDO

Relativity and quantum theory have shown that it has no meaning to divide the observing apparatus from what is observed.

DAVID BOHM

There is neither seer nor seeing nor seen. There is but one Reality—changeless, formless and absolute. How can it be divided?

SHANKARA

The concept of consciousness in fact demands a cut between subject and object, the existence of which is a logical necessity, which the position of the cut is to a certain extent arbitrary.

WOLFGANG PAULI

We generally distinguish between inner and outer, but...the distinction is no more than a form of thought construction.... Change the position, and what is inner is outer, and what is outer is inner.

D. T. SUZUKI

Without being aware of it and without being rigorously systematic about it, we exclude the Subject of Cognizance from the domain of nature that we endeavor to understand. We step with our own person back into the part of an onlooker who does not belong to the world, which by this very process becomes an objective world.

ERWIN SCHRÖDINGER

Failing to recognize me, you objectify me as an external entity.
But when you finally discover me,
The one naked mind arisen from within,
Absolute Awareness permeates the Universe.

YESHE TSOGYEL

In introspection it is clearly impossible to distinguish sharply between the phenomena themselves and their conscious perception.

NIELS BOHR

⁓

If the object did not have the nature of awareness, it would be without illumination, as it was before [its appearance]. Awareness cannot be different [than the object]. Awareness is the essential nature of the object.

UTPALADEVA

Name and Form

Modern physics has revealed a world that is made up not of elementary particles but of elementary ideas, forms and symmetries. Many of the properties previously associated with "matter" do not apply to atomic particles, suggesting to scientists that the world is fundamentally more akin to mind than to what we used to think of as matter. In fact, so many common-sense notions of reality have turned out to be wrong that physicists now view the external world as a mental construct—and one that can sometimes be misleading.

The world of the new physics bears surprising similarities to that of Eastern mysticism, which is said to consist of name and form. All things are essentially forms or ideas created by the act of naming, a habitual, unconscious process that normally takes place outside our awareness. In fact, it is this very lack of awareness that leads us to believe the world has its own reality independent of the mind. Like the mistaken view of classical physics, this delusion blinds us to the true nature of things.

Physical concepts are free creations of the human mind, and are not, however it may seem, uniquely determined by the external world.

ALBERT EINSTEIN

All such notions as causation, succession, atoms, primary elements...are all figments of the imagination and manifestations of the mind.

BUDDHA

Excepting immediate sensations and, more generally, the content of my consciousness, everything is a construct...but some constructs are closer, some farther, from the direct sensations.

EUGENE P. WIGNER

Apart from the Absolute, nothing else differs from name and form, since all modifications are but manifestations of name and form.

SHANKARA

The smallest units of matter are in fact not physical objects in the ordinary sense of the word; they are forms.

Werner Heisenberg

All things—from Brahma the creator down to a single blade of grass—are the apparently diverse names and forms of the one Atman.

Shankara

The universe can be best pictured...as consisting of pure thought.

SIR JAMES JEANS

In fact, what is called the world is only a thought.

SRI RAMANA MAHARSHI

Science is the attempt to make the chaotic diversity of our sense-experience correspond to a logically uniform system of thought.... The sense-experiences are the given subject-matter. But the theory that shall interpret them is man-made. It is...hypothetical, never completely final, always subject to question and doubt.

ALBERT EINSTEIN

The concept of matter or of material object arises only in the co-ordinating and interpreting mental consciousness.... Outer objects of our consciousness, including those which we call 'material' and which make up our apparently solid and tangible world, are real only in a relative sense.

LAMA ANAGARIKA GOVINDA

I do not think that sensation, as we know it, could exist without an activity of the mind which concentrates, compares and distinguishes. What we call a sensation can never be purely sensory.... The most primitive data we can reach will not be wholly independent of the primitive forms of thought.

SIR ARTHUR EDDINGTON

When we see a tree and call it a tree, we think this sense experience is final; but in point of fact this sense experience is possible only when it is conceptualized. A tree is not a tree until it is subsumed under the concept "tree." Tathata is what precedes this conceptualization; it is where we are even before we say it is or it is not.

D. T. SUZUKI

Every measurement first acquires its meaning for physical science through the significance which a theory gives it.

MAX PLANCK

Every existent object is a product of something to be given a name and something else to give it a name. There is not a single atom of anything in the universe which does not rely on this process—there is nothing that exists from its own side.

PABONGKA RINPOCHE

Even scholars of audacious spirit and fine instinct can be obstructed in the interpretation of facts by philosophical prejudices. The prejudice...consists in the faith that the facts by themselves can and should yield scientific knowledge without free conceptual construction. Such a misconception is possible only because one does not easily become aware of the free choice of such concepts, which, through verification and long usage, appear to be immediately connected with the empirical material.

ALBERT EINSTEIN

⸺

The main trouble with the human mind is that while it is capable of creating concepts in order to interpret reality it hypostatizes them and treats them as if they were real things. Not only that, the mind regards its self-constructed concepts as laws externally imposed upon reality, which has to obey them in order to unfold itself. This attitude or assumption on the part of the intellect helps the mind to handle [superficial] nature for its own purposes, but the mind altogether misses the [deep] inner workings of life and consequently is utterly unable to understand it.

D. T. SUZUKI

If the universe is a universe of thought, then its creation must have been an act of thought.

SIR JAMES JEANS

The entire universe and everything in it is conceptually designated. The sutras state that all these phenomena are designated by thoughts.

GEN LAMRIMPA

"Being" is always something which is mentally constructed by us, that is, something which we freely posit (in the logical sense). The justification of such...constructs, which represent "reality" for us, lies alone in their quality of making intelligible what is sensorily given.

ALBERT EINSTEIN

The mere fact that something is conceptually designated does not necessarily indicate that it exists.... In other words, if all it took for something to exist is that it is conceptually designated, then anything that comes to mind would exist.

GEN LAMRIMPA

Illusions and Delusions

Ancient physics demonstrated that the earth was round, not flat, and Copernicus dispelled the mistaken notion that the earth was the center of the universe. Similarly, modern physics has freed us from other illusions. Space and time are neither independent nor absolute; matter and energy are not ultimately distinct; and the attributes of objects do not exist independent of observation.

Eastern spiritual traditions also strive to free us from the belief that things and their relationships are "out there." The nature of our perceptions, they tell us, is dreamlike. When we recognize that we are dreaming, the illusion that the dream world is real vanishes—even if we continue to dream.

For the scientist, as long as knowledge of reality remains in the realm of theory, new layers of illusion always await unmasking. The mystic, however, strives not to replace one illusion with another, subtler one, but to achieve freedom from all illusions, forever.

Time and again the passion for understanding has led to the illusion that man is able to comprehend the objective world rationally by pure thought without any empirical foundations—in short, by metaphysics.

ALBERT EINSTEIN

By becoming attached to names and forms, not realizing that they have no more basis than the activities of the mind itself, error arises and the way to emancipation is blocked.

BUDDHA

It has been the task of science to discover that things are very different from what they seem.

SIR ARTHUR EDDINGTON

Analytical discernment can be used to realize that...[things] have no intrinsic reality. With this realization, the basis of deception collapses from its inner core.

KONGTRUL LODRÖ TAYÉ

The advance of physics has emancipated us from some of the common forms of thought. We use them, but we are not deceived by them.

SIR ARTHUR EDDINGTON

Being under illusion means perceiving objective appearances and mental appearances as having independent reality.... When illusion ceases, appearances continue to exist but they are no longer assimilated as objects grasped by a subject.

BOKAR RINPOCHE

In our thinking...we attribute to this concept of the bodily object a significance, which is to high degree independent of the sense impression which originally gives rise to it. This is what we mean when we attribute to the bodily object "a real existence...." By means of such concepts and mental relations between them, we are able to orient ourselves in the labyrinth of sense impressions. These notions and relations...appear to us as stronger and more unalterable than the individual sense experience itself, the character of which as anything other than the result of an illusion or hallucination is never completely guaranteed.

ALBERT EINSTEIN

I teach that the multitudinousness of objects have no reality in themselves but are only seen of the mind and, therefore, are of the nature of maya and a dream.... It is true that in one sense they are seen and discriminated by the senses as individualized objects; but in another sense, because of the absence of any characteristic marks of self-nature, they are not seen but are only imagined. In one sense they are graspable, but in another sense, they are not graspable.

BUDDHA

How, then, have they [scientific revolutions] come about?
The answer that comes readiest to hand...is because the
new ideas are simply right and the old ones wrong. This
answer presupposes that in science it is always the right
answer that prevails. But that is by no means the case. For
example, the correct notion of a heliocentric planetary
system, developed by Aristarchus, was abandoned in favor
of Ptolemy's geocentric viewpoint, although the latter was
false.... One cannot be absolutely certain that the right
theory will always triumph.

WERNER HEISENBERG

Once Chuang Chou dreamt he was a butterfly.... Suddenly
he woke up and there he was, solid and unmistakable
Chuang Chou. But he didn't know if he was Chuang Chou
who had dreamt he was a butterfly, or a butterfly dreaming
he was Chuang Chou.

CHUANG TZU

The external world of physics has thus become a world of shadows. In removing our illusions we have removed the substance, for indeed we have seen that substance is one of the greatest of our illusions.

SIR ARTHUR EDDINGTON

Although not really existing, things still appear. From their own side, however, (such things) are void by nature. These void appearances do not actually exist.... They have no foundation, no support, no beginning, middle or end.

LONGCHEN RABJAMPA (LONGCHENPA)

We have torn away the mental fancies to get at the reality beneath, only to find that the reality of that which is beneath is bound up with its potentiality of awakening these fancies. It is because the mind, the weaver of illusion, is also the only guarantor of reality that reality is always to be sought at the base of illusion. Illusion is to reality as the smoke to the fire.

SIR ARTHUR EDDINGTON

That which is presupposed by the power to imagine, that from which imagination proceeds, cannot itself be imagined.... That from which all imaginations such as "real" and "unreal" proceed, together with the power to think and reflect, is itself non-dual and (in the highest sense) real.

SHANKARA

All earlier systems of physics...fell into the error of identifying appearance with reality; they confined their attention to the walls of the cave, without even being conscious of a deeper reality beyond. The new quantum theory has shown that we must probe the deeper substratum of reality before we can understand the world of appearance.

SIR JAMES JEANS

What we see here displayed or performed are passing shadows of something behind, and...when the latter is not finally grasped by our experience the meaning of the passing shadows will never be properly recognized and appraised.

D. T. SUZUKI

No single man can make a distinction between the realm
of his perceptions and the realm of things that cause it
since, however detailed the knowledge he may have
acquired about the whole story, the story is occurring only
once not twice. The duplication is only an allegory,
suggested mainly by communication with other human
beings and even with animals; which shows that their
perceptions in the same situation seem to be very similar
to his own apart from insignificant differences in the point
of view.

ERWIN SCHRÖDINGER

No matter what a deluded man may think he is perceiving,
he is really seeing Brahman and nothing else but Brahman.
He sees mother-of-pearl and imagines that it is silver. He
sees Brahman and imagines that it is the universe.

SHANKARA

Waves, Fields and Energy

According to Einstein's most famous equation, $E=mc^2$, the mass of a particle is equivalent to the energy of a wave. Quantum physics describes matter as waves of probability; the fundamental nature of matter is that of a wave or ripple on a vast ocean of energy. Only at the brief moment of observation does it have the properties of a particle. The modern physicist views a particle as simply a small portion of wave energy that has temporarily condensed into a localized mass and will inevitably dissolve again into the sea of energy from which it came.

The mystic's insight into the nature of phenomena also reveals a realm of subtle energetic vibrations behind the appearances of physical objects. The world of appearances is merely a play of waves on the surface of an infinite ocean of consciousness. Just as waves and the sea are both made of water, so the world and consciousness are ultimately the same substance.

We find that we can best understand the course of events in terms of waves of knowledge.

SIR JAMES JEANS

Mental formations with regard to objects and mind agitate the fundamental consciousness like waves on water.

KONGTRUL LODRÖ TAYÉ

We may picture the world of reality as a deep-flowing stream; the world of appearance is its surface, below which we cannot see. Events deep down in the stream throw up bubbles and eddies on the surface of the stream. These are the transfers of energy and radiation of our common life, which affect our senses and so activate our minds; below these lie deep waters.

Sir James Jeans

The wave, the foam, the eddy and the bubble are all essentially water. Similarly, the body and the ego are really nothing but pure consciousness. Everything is essentially consciousness.

Shankara

Matter is like a small ripple on this tremendous ocean of energy, having some relative stability and being manifest.... And in fact beyond that ocean may be still a bigger ocean...the ultimate source is immeasurable and cannot be captured within our knowledge.

DAVID BOHM

Universal Mind is like a great ocean, its surface ruffled by waves and surges but its depth remaining forever unmoved.

BUDDHA

The field theory of matter...[states that] a material particle such as an electron is merely a small domain of the electrical field within which the field strength assumes enormously high values, indicating that a comparatively huge field energy is concentrated in a very small space. Such an energy knot, which by no means is clearly delineated against the remaining field, propagates through empty space like a water wave across the surface of a lake; there is no such thing as one and the same substance of which the electron consists at all times.

HERMANN WEYL

Such products of water as foam, ripples, waves and bubbles, though non-different from the water of the sea, are mutually distinct and join one another and enter into other such mutual relations. And though they are non-different from the water of the sea, these products of water, foam, waves and the rest, are not mutually identical. And though they are not mutually identical, this does not affect the fact that they are non-different from the sea. In the same way, in the case under consideration, the experiencer and the objects of his experience need not be mutually identical, though they remain non-different from the Absolute.

SHANKARA

There is no essential distinction between mass and energy. Energy has mass and mass represents energy. Instead of two conservation laws we have only one, that of mass-energy.

ALBERT EINSTEIN

...Only an arbitrary distinction in thought divides form of substance from form of energy.

Matter expresses itself eventually as a formulation of some unknown Force.

SRI AUROBINDO

We could regard matter as the regions in space where the field is extremely strong.... There would be no place, in our new physics, for both field and matter, field being the only reality.

ALBERT EINSTEIN

In contrast to space is the principle of substance, of differentiation, of "thingness." But nothing can exist without space. Space is the precondition of all that exists, be it in material or immaterial form, because we can neither imagine an object nor a being without space.

LAMA ANAGARIKA GOVINDA

The elementary particles...do not themselves consist of matter, but they are the only possible forms of matter. Energy becomes matter by taking on the form of an elementary particle, by manifesting itself in this form.

WERNER HEISENBERG

In a certain sense Matter is unreal and non-existent; that is to say, our present knowledge, idea and experience of Matter is not its truth, but merely a phenomenon of particular relation between our senses and the all-existence in which we move. When Science discovers that Matter resolves itself into forms of Energy, it has hold of a universal and fundamental truth.

SRI AUROBINDO

Particles and Matter

When Erwin Schrödinger published his quantum wave equation in 1926, it swept away the notion of classical physics that each atom was a kind of microscopic solar system in which electrons orbited around a nucleus. Separate, localized particles of matter dissolved into nonseparable, nonlocal waves. To make the new physics even more puzzling, scientists soon realized that the waves were not vibrations of some substance, like sounds in the air, but rather waves of probability. The classical idea of material existence was exposed as an illusion.

In Eastern mysticism, too, all objects are recognized as illusory, empty of any independent existence. Their fundamental reality is not material but spiritual. A dualism of matter versus spirit arises from the mistaken belief that things have their own independent existence. As spiritual knowledge deepens, illusions of materiality and duality are exposed, revealing the ineffable identity of form and emptiness.

Atoms are not things.

WERNER HEISENBERG

⸺ ⸺

For the wise all "things" are wiped away.

BUDDHA

The concept of substance has disappeared from
fundamental physics.

Sir Arthur Eddington

I have seen nothing in the world that is ultimately real.

Yeshe Tsogyel

The external world...is not a thing founded in itself, that can in a significant manner be established as an independent existence.

HERMANN WEYL

There is nothing that can be called an independent, solitary, self-originating primary nature. All is ultimately empty, and if there is such a thing as primary nature, it cannot be otherwise than empty.

D. T. SUZUKI

The materialist must decide for himself...whether the ghostly remains of matter should be labeled as matter or something else; it is mainly a question of terminology. What remains is in any case very different from the full-blooded matter and the forbidding materialism of the Victorian scientist.

SIR JAMES JEANS

But there is that which does not belong to materialism and which is not reached by the knowledge of the philosophers who cling to false discriminations and erroneous reasonings because they fail to see that, fundamentally, there is no reality in external objects.

BUDDHA

The atoms or the elementary particles...form a world of potentialities or possibilities rather than one of things or facts.

WERNER HEISENBERG

In Buddhist Emptiness there is no time, no space, no becoming, no-thing-ness; it is what makes all things possible; it is a zero full of infinite possibilities, it is a void of inexhaustible contents.

D. T. SUZUKI

That environment of space and time and matter, of light and colour and concrete things, which seems so vividly real to us...has melted into a shadow.

SIR ARTHUR EDDINGTON

All the mind's arbitrary conceptions of matter, phenomena, and of all conditioning factors and all conceptions and ideas relating thereto are like a dream, a phantasm, a bubble, a shadow.

BUDDHA

Our conception of substance is only vivid so long as we do not face it. It begins to fade when we analyze it. We may dismiss many of its supposed attributes which are evidently projections of our sense-impressions outwards into the external world.

SIR ARTHUR EDDINGTON

Ignorance apprehends its object as if it exists purely objectively.... If phenomena in fact existed in this way...then the more carefully one investigated them, the more clearly those objects would appear to the mind. However, the more closely one examines phenomena, the more one sees that objects are not to be found under such analysis.

GEN LAMRIMPA

Quantum theory requires us to give up the idea that the electron, or any other object has, by itself, any intrinsic properties at all.

DAVID BOHM

By emptiness of self-aspect or self-character, therefore, is meant that each particular object has no permanent and irreducible characteristics to be known as its own.

D. T. SUZUKI

Practical realism has always been and will always be an essential part of natural science. Dogmatic realism, however, is...not a necessary condition for natural science.... Actually the position of classical physics is that of dogmatic realism. It is only through quantum theory that we have learned that exact science is possible without the basis of dogmatic realism.

WERNER HEISENBERG

When you do this sort of analysis, and you seek the thing with the name, you will never be able to find a single atom of anything in the universe that exists in itself. All the normal workings of the world though are quite logical and proper; things make other things happen, things do what they do, though all in only an apparent way, and a conventionally agreed-upon way.

PABONGKA RINPOCHE

Wholeness and Interdependence

Quantum physics reveals the world to be an intricate web of events that are interrelated in mysterious, previously unimaginable ways. Instantaneous correlations between distant particles show that despite appearances, nature is an inseparable whole, not merely a collection of parts. Elementary particles can be transformed into each other, as well as into pure energy. In a certain sense, each particle implicitly contains all others. The whole spectrum of particles represents the set of possible states of a single matter field.

Mystical traditions, too, teach the fundamental unity of all things. Worldly objects and events derive from an ultimate wholeness of which they are interdependent, interrelated aspects. The world, as Eastern mystics tell us, consists of harmonious relationships between parts that are ultimately identical with their common source, forming a coherent and ordered whole that is strikingly similar in concept to the realm of quantum physics.

The elementary particles are certainly not eternal and indestructible units of matter, they can actually be transformed into each other.

WERNER HEISENBERG

Objects are arrayed in such a way that their mutual separateness no more exists.... There is here a state of interpenetration of all objects.

D. T. SUZUKI

Non-separability...shows that the general approach of Democritian atomism is a false view of nature even when applied to events. If we call "atoms" micro-objects, having definite properties, or micro-events, then it is we who, so to speak, paint the distinct atoms on the canvas of non-separable reality, whatever this latter word means.

BERNARD D'ESPAGNAT

The Way has never known boundaries.... But because of the recognition of a "this," there came to be boundaries.

CHUANG TZU

If our small minds, for some convenience, divide this...universe into parts—physics, biology, geology, astronomy, psychology, and so on—remember that nature does not know it! So let us put it all back together, not forgetting ultimately what it is for. Let it give us one more final pleasure: drink it and forget it all!

RICHARD P. FEYNMAN

Forget distinctions. Leap into the boundless and make it your home!

CHUANG TZU

The experimentally known fact that all the elementary particles can be transformed into one another is an indication that it could scarcely be possible to single out one particular group of such particles.... One finds structures [in the theory of elementary particles] so linked and entangled with each other that it is really impossible to make further changes at any point without calling all the connections into question.

WERNER HEISENBERG

Each phenomenon is determining every other phenomenon and is simultaneously being determined by each and every phenomenon. This feature of mutual determinacy, or interdependency, of all phenomena is sometimes translated as mutual identity. Moreover, according to this doctrine, not only are all phenomena interdependent, but they also interpenetrate without any hindrance.

CHENG CHIEN

Light is what enfolds all the universe.... Light in its generalized sense (not just ordinary light) is the means by which the entire universe unfolds into itself.

DAVID BOHM

Individual realities are enfolded in one great Reality—a world of lights not accompanied by any form of shade. The essential nature of light is to intermingle without interfering or obstructing or destroying one another. One single light reflects in itself all other lights generally and individually.

D. T. SUZUKI

Every particle consists of all other particles.... We could not say that the proton consists of three quarks. We would have to say that it may temporarily consist of three quarks, but may also temporarily consist of four quarks and one antiquark, or five quarks and two antiquarks, and so on.

WERNER HEISENBERG

Every phenomenon "contains" every other phenomenon, and every phenomenon also "contains" the totality of all phenomena which interpenetrate in perfect freedom and non-obstruction.

CHENG CHIEN

An object does not have any "intrinsic" properties (for instance, wave or particle) belonging to itself alone; instead, it shares all its properties mutually and indivisibly with the systems with which it interacts.

DAVID BOHM

Things derive their being and nature by mutual dependence and are nothing in themselves.

SIDDHA NAGARJUNA

The world thus appears as a complicated tissue of events, in which connections of different kinds alternate or overlap or combine and thereby determine the texture of the whole.

WERNER HEISENBERG

The external world and his inner world are for him [the Buddhist] only two sides of the same fabric, in which the threads of all forces and all events, of all forms of consciousness and of their objects, are woven into an inseparable net of endless, mutually conditioned relations.

LAMA ANAGARIKA GOVINDA

What is needed is for man to give attention to his habit of fragmentary thought, to be aware of it, and thus bring it to an end. Man's approach to reality may then be whole, and so the response will be whole.

DAVID BOHM

To come directly into harmony with this reality just simply say when doubt arises, "Not two." In this "not two" nothing is separate, nothing is excluded.

SENGTSAN

Time and Space

Einstein called conventional beliefs about time and space "stubborn, persistent illusions." His Special Theory of Relativity established that time and space, as separate, objective realities simply do not exist. Instead, events take place in a unified, four-dimensional space-time continuum.

Each of us sees a different projection of the continuum, as if observing the shadows cast by a tree when the sun is in different positions. Before Einstein, science took the shadow as the objective reality and the ground as the space containing it. Now we know the shadow is only an appearance created by a larger reality that includes the sun and the tree.

Without the benefit of Einstein's mathematical equations, Eastern contemplatives have understood for centuries that time and space are indivisibly linked—and ultimately not real. In the deepest states of mystical insight, time and space are transcended altogether, revealing them, and Einstein's time-space continuum as well, to be illusory.

People like us, who believe in physics, know that the distinction between past, present and future is only a stubborn, persistent illusion.

ALBERT EINSTEIN

The past, the future...are nothing but names, forms of thought, words of common usage, merely superficial realities.

T. R. V. MURTI

The common words "space" and "time" refer to a structure of space and time that is actually an idealization and oversimplification.

WERNER HEISENBERG

There is nothing like an absolute time which remains as a reality apart from successive events. Time and space are derived notions, modes of reference.

K. VENKATA RAMANAN

This requirement [of general relativity] ...takes away from space and time the last remnant of physical objectivity.

ALBERT EINSTEIN

Space is not an entity.

SIDDHA NAGARJUNA

Eternally and always there is only now, one and the same now; the present is the only thing that has no end.

Erwin Schrödinger

The past and future are both rolled up in this present moment of illumination, and this present moment is not something standing still with all its contents, for it ceaselessly moves on.

D. T. Suzuki

What we mean by "right now" is a mysterious thing which we cannot define.... "Now" is an idea or concept of our mind; it is not something that is really definable physically at the moment.

RICHARD P. FEYNMAN

Words! The way is beyond language, for in it there is no yesterday, no tomorrow, no today.

SENGTSAN

The theory of relativity [showed] that even such fundamental concepts as space and time could be changed and in fact must be changed on account of new experience.

WERNER HEISENBERG

⸻

Our consciousness determines the kind of space in which we live.... The way in which we experience space, or in which we are aware of space, is characteristic of the dimension of our consciousness.

LAMA ANAGARIKA GOVINDA

The four-dimensional space-time manifold is only a fabrication, only a theory.

JOHN A. WHEELER

Space and time...are names.

SIDDHA NAGARJUNA

Henceforth space by itself, and time by itself, are doomed to fade away into mere shadows, and only a kind of union of the two will preserve an independent reality.

H. Minkowski

As a fact of pure experience, there is no space without time, nor time without space; they are interpenetrating.

D. T. Suzuki

The concepts of happening and becoming are indeed not completely suspended, but yet complicated. It appears therefore more natural to think of physical reality as a four-dimensional [space-time] existence, instead of, as hitherto, the [temporal] evolution of a three-dimensional [spatial] existence.

ALBERT EINSTEIN

In this space-experience the temporal sequence is converted into a simultaneous co-existence...and this again does not remain static but becomes a living continuum in which time and space are integrated.

LAMA ANAGARIKA GOVINDA

What we perceive through the senses as empty space...is the ground for the existence of everything, including ourselves. The things that appear to our senses are derivative forms and their true meaning can be seen only when we consider the plenum, in which they are generated and sustained, and into which they must ultimately vanish.

DAVID BOHM

Wherefrom do all these worlds come? They come from space. All beings arise from space, and into space they return: space is indeed their beginning, and space is their final end.

THE UPANISHADS

According to general relativity, the concept of space detached from any physical content does not exist.

ALBERT EINSTEIN

If there is only empty space, with no suns nor planets in it, then space loses its substantiality.

BUDDHA

Manifestation and Causality

Modern physics reveals the world to be a hierarchy of levels. Concrete planes of existence make up the more superficial levels, while at deeper levels are subtler planes of reality where things exist in a unified state of potentiality. At different levels, different degrees of cause and effect apply. Quantum theory introduced a fundamental element of indeterminism into physics. Spontaneity, we learned, is intrinsic to nature on the most basic level. Cause and effect still apply, but only to the probability of events, not the events themselves.

Eastern mystical traditions also view reality as a hierarchy of levels of manifestation, from gross levels where objects appear to have their own existence, to subtle levels of energetic vibrations that transcend ordinary space and time. Like causality in physics, the law of karma links actions with results. Yet although karma never fails within the plane of space and time, it is not absolute: at the ultimate level of reality, there are spontaneity, freedom and grace.

The more a man is imbued with the ordered regularity of all events the firmer becomes his conviction that there is no room left by the side of this ordered regularity for causes of a different nature. For him neither the rule of human nor the rule of divine will exists as an independent cause of natural events.

ALBERT EINSTEIN

A person's entered the path that pleases Buddhas
When for all objects, in the cycle and beyond,
He sees that cause and effect can never fail,
And when for him they lose all solid appearance.

TSONGKAPA

We have been forced step by step to forego a causal description of the behavior of individual atoms in space and time, and to reckon with a free choice on the part of nature between various possibilities.

NIELS BOHR

The doctrine of causation...is only applicable to a world of dualities and combinations. Where there are no such happenings, the doctrine at once loses its significance. As long as we are bound to a world of particulars we see causation and relativity everywhere, because this is the place for them to function.... The realm of Emptiness...lies underneath the world of causes and conditions. Emptiness is that which makes the work of causation possible.

D. T. SUZUKI

Causality may be considered as a mode of perception by which we reduce our sense impressions to order.

NIELS BOHR

～～

Time, space, and causation are like the glass through which the Absolute is seen.... In the Absolute there is neither time, space, nor causation.

VIVEKANANDA

Everything we know about Nature is in accord with the idea that the fundamental process of Nature lies outside space-time...but generates events that can be located in space-time.

H. P. STAPP

Parama Shiva is beyond the limits of time, space and form; and as such is eternal and infinite.... His nature has primarily a two-fold aspect—an imminent aspect in which He pervades the universe, and a transcendent aspect in which He is beyond all Universal Manifestation.

JAGADISH CHANDRA CHATTERJI

At any given stage of the theoretical construction [in science] there exists a hierarchy of laws, inasmuch as different degrees of stability are ascribed to the different laws.... Thus in the practice of scientific research the clear-cut division into a priori and a posteriori in the Kantian sense is absent, and in its place we have a rich scale of gradations of stability.

HERMANN WEYL

The order of dissolution must the be exact reverse of the order of emergence.... One must reckon that each successive cause in the hierarchy of causes among the elements dissolves back into its own cause, the next and more subtle principle in the series, until the whole mass of effects dissolves finally into the Absolute, the highest and most subtle cause of all.

SHANKARA

Whatever may be the nature of these inward depths of consciousness, they are the very ground, both of the explicit content and of that content which is usually called implicit. Although this ground may not appear in ordinary consciousness, it may nevertheless be present in a certain way.

DAVID BOHM

The ultimate creative principle is consciousness. There are different levels of consciousness. What we call innermost subtle consciousness is always there.... All of our other [kinds of] consciousnesses—sense consciousness and so on—arise in dependence on this mind of clear light.

THE DALAI LAMA

Quantum mechanics suggests that this is the way that phenomenal reality comes about from a deeper order in which it is enfolded. Reality unfolds to produce the visible order and folds back in. It is constantly unfolding and enfolding.

DAVID BOHM

In the structure of Buddhist thought as well as in the way that it is expounded are found two activities or movements, one of "going forth" or "going up-ward," the other of "coming back," or "coming down." The two activities for the sake of convenience can be named simply "ascent" and "descent."

GADJIN M. NAGAO

The Universe had to have a way to come into being out of nothingness.... When we say "out of nothingness" we do not mean out of the vacuum of physics. The vacuum of physics is loaded with geometrical structure and vacuum fluctuations and virtual pairs of particles. The Universe is already in existence when we have such a vacuum. No, when we speak of nothingness we mean nothingness: neither structure, nor law, nor plan.... For producing everything our of nothing one principle is enough.

JOHN A. WHEELER

In order to create a universe, then, Parama Shiva brings into operation that aspect of his Shakti which manifests itself as the principle of Negation and lets the ideal Universe disappear from His view and allows Himself, as it were, to feel the want of a Universe, but for which feeling there could be...no need of a manifested Universe on the part of one who is all-complete in Himself.

JAGADISH CHANDRA CHATTERJI

Unity and Plurality

For the physicist, understanding means discovering the universal principles of nature. Newton's law of gravitation unified in one equation the motion of planets and the behavior of objects falling on earth. As physics deepens, the laws become more universal. Just as in the 17th century Newton unified the terrestrial and celestial laws, in the 19th century Maxwell unified the laws of electricity and magnetism, and 20th-century physicists unified the electromagnetic laws with those of the weak nuclear force. The holy grail of 21st-century physics is to unify the laws governing all known forces of nature within a single "theory of everything."

For the mystic, too, the holy grail is to discover the unity behind all things. The goal, however, is not a theory but transcendence of knowledge in all its forms. The mystic's unity embraces not only all phenomena of the world but his or her own self as well. Like the physicist, the mystic seeks to discover the invariant within all variation, the One at the root of the Many.

We think of a substance as something which can be neither created nor destroyed.

ALBERT EINSTEIN

He is the Eternal among things that pass away.

THE UPANISHADS

We know that there is an ever-changing variety of phenomena appearing to our senses. Yet we believe that ultimately it should be possible to trace them back somehow to some one principle.

WERNER HEISENBERG

The Self, the unborn indestructible principle of reality, undergoes modification through illusion (maya) only and not in the real sense. Hence duality is not in the ultimate sense real.

SHANKARA

It is in this striving after the rational unification of the manifold that it encounters its greatest successes, even though it is precisely this attempt which causes it to run the greatest risk of falling a prey to illusions.

ALBERT EINSTEIN

If there is even a trace of this and that, of right and wrong, the Mind-essence will be lost in confusion. Although all dualities come from the One, do not be attached even to this One.

SENGTSAN

A theory is the more impressive the greater the simplicity of its premises is, the more different kinds of things it relates, and the more extended is its area of applicability.

ALBERT EINSTEIN

As in Science, so in metaphysical thought, that general and ultimate solution is likely to be the best which includes and accounts for all so that each truth of experience takes its place in the whole.

SRI AUROBINDO

We always look for the basic thing behind the dependent thing, for what is absolute behind what is relative, for the reality behind the appearance and for what abides behind what is transitory. In my opinion, this is characteristic not only of physical science but of all science.

MAX PLANCK

The means to be depended upon to know the nonduality which is Ultimate Reality is nothing but cognizing the nondiversity in the diversity of the manifestation.

ABHINAVAGUPTA

To be confused about what is different and what is not, is
to be confused about everything.

DAVID BOHM

⸺ ⸺

There are not many but only One.
Who sees variety and not the unity wanders from death to
death.

THE UPANISHADS

The plurality that we perceive is only an appearance; it is not real.

ERWIN SCHRÖDINGER

Whatever you see as duality is unreal.

SHANKARA

The world is given but once. Nothing is reflected. The original and the mirror-image are identical. The world extended in time and space is but our representation. Experience does not give us the slightest clue of its being anything besides that.

ERWIN SCHRÖDINGER

The fact is, there is only one world—there are not two worlds.... People think there are two worlds by the activity of their own minds. If they could get rid of these false judgements and keep their minds pure with the light of wisdom, then they would see only one world and that world bathed in the light of wisdom.

BUDDHA

Physics and Mysticism

"All religions, arts and sciences are branches of the same tree," Einstein wrote. "All these aspirations are directed toward ennobling man's life, lifting it from the sphere of mere physical existence and leading the individual toward freedom."

Modern physicists and modern mystics have commented on the similarities between science and religion, as well as their differences. In the first part of the 20th century, when the differences appeared greater than ever before, the discoverers of quantum theory noticed striking similarities with certain ideas of Eastern religions.

Contemporary authorities of Eastern religions, such as the Dalai Lama and Sri Aurobindo, are familiar with the substance of modern physics and have commented on its relationship to religion. Moving into the 21st century, we can expect that more physicists will become familiar with the world's mystical traditions, as well. Perhaps we can even hope that they will help lead the way to a new era of harmonious integration between science and religion.

Even though the realms of religion and science in themselves are clearly marked off from each other, nevertheless there exist between the two strong reciprocal relationships and dependencies.... The situation may be expressed by an image: Science without religion is lame, religion without science is blind.

ALBERT EINSTEIN

True reconciliation proceeds always by a mutual comprehension leading to some sort of intimate oneness. It is therefore through the utmost possible unification of Spirit and Matter that we shall best arrive at their reconciling truth and so at some strongest foundation for a reconciling practice in the inner life of the individual and his outer existence.

SRI AUROBINDO

The general notions about human understanding...which are illustrated by discoveries in atomic physics are not in the nature of things wholly unfamiliar, wholly unheard of, or new. Even in our own culture they have a history, and in Buddhist and Hindu thought a more considerable and central place. What we shall find is an exemplification, an encouragement, and a refinement of old wisdom.

J. ROBERT OPPENHEIMER

There is a growing interest among the scientific community in Buddhist philosophical thought. I am optimistic that over the next few decades there will be a great change in our worldview both from the material and the spiritual perspectives.

THE DALAI LAMA

Many people think that modern science is far removed from God. I find, on the contrary, that...in our knowledge of physical nature we have penetrated so far that we can obtain a vision of the flawless harmony which is in conformity with sublime reason.

HERMANN WEYL

Science at its limits, even physical science, is compelled to perceive in the end the infinite, the universal, the spirit, the divine intelligence and will in the material universe.

SRI AUROBINDO

Science demands also the believing spirit. Anybody who has been seriously engaged in scientific work of any kind realizes that over the entrance to the gates of the temple of science are written the words: Ye must have faith. It is a quality which the scientists cannot dispense with.

MAX PLANCK

Faith is the only entrance to Buddhism. Without faith all earnest study and constant effort will be of no avail. Just as soon as you are convinced that error always follows doubt, give up all doubt and enter the gateway of the faith.

CHIH-CHI

Physics is reflection on the divine Ideas of Creation, therefore physics is divine service. We are in our time very far from this theological foundation or justification of physics; but we still follow this method, because it has been so successful.

WERNER HEISENBERG

Matter reveals itself to the realizing thought and to the subtilized senses as the figure and body of Spirit....

In the light of this conception we can perceive the possibility of a divine life for man in the world which will at once justify Science by disclosing a living sense and intelligible aim for the cosmic and the terrestrial evolution....

SRI AUROBINDO

Modern science may have brought us closer to a more satisfying conception of this relationship [between spirit and matter] by setting up, within the field of physics, the concept of complementarity. It would be most satisfactory of all if physics and psyche could be seen as complementary aspects of the same reality.

WOLFGANG PAULI

The two are one: Spirit is the soul and reality of that which we sense as Matter; Matter is a form and body of that which we realize as Spirit.

SRI AUROBINDO

We shall not expect the natural sciences to give us direct insight into the nature of the spirit.

ERWIN SCHRÖDINGER

Spiritual truth is a truth of the spirit, not a truth of the intellect, not a mathematical theorem or a logical formula.

SRI AUROBINDO

Bibliography

In the citations below, the first page number refers to *Einstein and Buddha*, and the second page number refers to the edition cited.

Abhinavagupta
Lawrence, David Peter. *Rediscovering God with Transcendental Argument*. (Albany, NY: SUNY Press, 1999).
  p. 152: p. 97.

Ashvaghosha
Goddard, Dwight, ed. *A Buddhist Bible*. (Boston: Beacon Press, 1970).
  p. 65: p. 372.

Aurobindo, Sri
Aurobindo, Sri. *The Life Divine*. (Pondicherry, India: Sri Aurobindo Ashram, 1970).
  p. 38: p. 12. / p. 51: p. 640. / p. 69: p. 362. / p. 102: p. 14. / p. 104: p. 234. / p. 151: p. 468. / p. 158: p. 25. / p. 162: p. 26. / p. 163: p. 241. / p. 164: p. 886.
Aurobindo, Sri. *The Synthesis of Yoga*. (Silver Lake, WI: Lotus Light Publications, 1992).
  p. 160: p. 492.

The Bhagavad Gita
The Bhagavad Gita. Mascaro, Juan, tr. (New York: Penguin, 1962).
  p. 27: p. 97. / p. 44: p. 100.

Bohm, David
Bohm, David. *Quantum Theory*. (New York: Dover, 1989).
  p. 61: p. 170. / p. 113: p. 139. / p. 122: p. 161.

Bohm, David. *Wholeness and the Implicate Order.* (New York: Routledge & Kegan Paul, 1983).

p. 70: p. 143. / p. 124: p. 7. / p. 135: p. 191-192. / p. 144: p. 210. / p. 153: p. 16.

Davies, P. C. W., ed. *The Ghost in the Atom: A Discussion of the Mysteries of Quantum Physics.* (Cambridge: Cambridge University Press, 1986).

p. 144: p. 121.

Weber, Renee. *Dialogues with Scientists and Sages: The Search for Unity.* (London: Routledge & Kegan Paul, 1986).

p. 68: p. 151. / p. 100: p. 28. / p. 120: p. 46.

Bohr, Niels

Bohr, Niels. *The Philosophical Writings of Niels Bohr, vol. I.* (Woodbridge, CT: Ox Bow Press, 1987).

p. 67: p. 54. / p. 139: p. 4. / p. 140: p. 116-117.

Bohr, Niels. *The Philosophical Writings of Niels Bohr, vol. II.* (Woodbridge, CT: Ox Bow Press, 1987).

p. 60: p. 59. / p. 73: p. 27.

Heisenberg, Werner. *Physics and Beyond.* (New York: Harper & Row, 1971).

p. 32: p. 206. / p. 45: p. 210. / p. 51: p. 209. / p. 54: p. 41. / p. 58: p. 40.

Bokar Rinpoche

Bokar Rinpoche. *Death and the Art of Dying in Tibetan Buddhism.* (San Francisco: Clear Point Press, 1993).

p. 90: p. 48. / p. 67: p. 48.

Buddha

The Dhammapada. Müller, Max F., tr. (New York: Colonial Press, 1902).

p: 24: ch. 24, verse 10 (line no. 343).

Goddard, Dwight, ed. *A Buddhist Bible.* (Boston: Beacon Press, 1970).

p. 36: p. 293. / p. 39: p. 320. / p. 40: p. 315. / p. 45: p. 310. / p. 50: p. 292. / p. 54: p. 104. / p. 60: p. 311. / p. 62: p. 106. / p. 64: p. 283. / p. 76: p. 314. / p. 88: p. 308. / p. 91: p. 297. / p. 100: p. 306. / p. 106: p. 302. / p. 109: p. 313. / p. 111: p. 102. / p. 136: p. 199. / p. 155: p. 632.

Chatterji, Jagadish Chandra
Chatterji, Jagadish Chandra. *Kashmir Shaivism.* (Albany, NY: SUNY Press, 1986).

    p. 141: p. 43-45. / p. 145: p. 64.

Chien, Cheng
Manifestation of the Tathagata. Chien, Cheng, tr. (Boston: Wisdom Publications, 1993).

    p. 119: p. 39. / p. 121: p. 39.

Chih-chi
Goddard, Dwight, ed. *A Buddhist Bible.* (Boston: Beacon Press, 1970).

    p. 161: p. 454.

The Dalai Lama
Gyatso, Tenzin, H. H. 14th Dalai Lama. *A Policy of Kindness.* (Ithaca, NY: Snow Lion, 1990).

    p. 26: p. 112. / p. 159: p. 71-72.
Weber, Renee. *Dialogues with Scientists and Sages: The Search for Unity.* (London: Routledge & Kegan Paul, 1986).

    p. 143: p. 237.

d'Espagnat, Bernard
d'Espagnat, Bernard. "Quantum Logic and Non-Separability," *The Physicist's Conception of Nature.* Mehra, Jagdish, ed. (Boston: D. Reidel Publishing Company, 1973).

    p. 117: p. 734.

Eddington, Sir Arthur
Eddington, Sir Arthur. *The Nature of the Physical World.* (Ann Arbor: University of Michigan Press, 1958).

    p. 39: p. 281. / p. 41: p. 333-334. / p. 65: p. 332. / p. 89: p. 334-335. / p. 93: p. xvi. / p. 94: p. 319. / p. 112: p. 273.
Eddington, Sir Arthur. *The Philosophy of Physical Science.* (Ann Arbor: University of Michigan Press, 1958).

    p. 36: p. 195. / p. 46: p. 155. / p. 81: p. 195. / p. 90: p. 118. / p. 107: p. 110.

Eddington, Sir Arthur. *Science and the Unseen World.* (New York: The MacMillan Company, 1929).

    p. III: p. 37-38.

Einstein, Albert

Einstein, Albert. *The Evolution of Physics.* (New York: Simon and Schuster, 1938).

    p. 76: p. 33. / p. 102: p. 208. / p. 148: p. 43. / p. 103: p. 258.

Einstein, Albert. *The Expanded Quotable Einstein.* Calaprice, Alice, ed. (Princeton: Princeton University Press, 2000).

    p. 24: p. 292. / p. 31: p. 116. / p. 26: p. 316. / p. 56: p. 261. / p. 126: p. 75.

Einstein, Albert. *Ideas and Opinions.* (New York: Crown Publishers, 1954).

    p. 28: p. 357. / p. 62: p. 266. / p. 88: p. 342. / p. 134: p. 371. / p. 136: p. 348. / p. 138: p. 48.

Einstein, Albert. *Out of My Later Years.* (New York: Philosophical Library, 1956).

    p. 27: p. 21. / p. 34: p. 115. / p. 80: p. 98. / p. 91: p. 60-61. / p. 150: p. 27. / p. 158: p. 24.

Einstein, Albert. *The Principle of Relativity.* (New York: Dover, 1952).

    p. 128: p. 117.

Einstein, Albert. *The World As I See It.* (Secaucus, NJ: Carol Publishing Group, 1999).

    p. 25: p. 7-8.

Schilpp, P. A., ed. *Albert Einstein: Philosopher-Scientist.* (Peru, IL: Open Court, 1970).

    p. 83: p. 49. / p. 85: p. 669. / p. 151: p. 33.

Feynman, Richard P.

Feynman, Richard P. *The Character of Physical Law.* (Cambridge, MA: M.I.T. Press, 1967).

    p. 38: p. 158. / p. 47: p. 128.

Feynman, Richard P. *The Feynman Lectures in Physics, vol. I.* (Reading, MA: Addison-Wesley, 1966).

    p. 118: p. 3-10. / p. 130: p. 17-4.

Govinda, Lama Anagarika

Govinda, Lama Anagarika. *Foundations of Tibetan Mysticism.* (York Beach, ME: Samuel Weiser, 1969).

p. 80: p. 68. / p. 103: p. 116. / p. 123: p. 93. / p. 131: p. 116. / p. 134: p. 116.

Heisenberg, Werner

Heisenberg, Werner. *Across the Frontiers.* (Woodbridge, CT: Ox Bow Press, 1990).

p. 40: p. 141. / p. 78: p. 116-117. / p. 92: p. 162-163. / p. 104: p. 22. / p. 119: p. 28. / p. 149: p. 105-106.

Heisenberg, Werner. *Encounters with Einstein.* (Princeton: Princeton University Press, 1989).

p. 121: p. 35. / p. 162: p. 9.

Heisenberg, Werner. *Physics and Beyond.* (New York: Harper & Row, 1971).

p. 37: p. 216. / p. 106: p. 11.

Heisenberg, Werner. *Physics and Philosophy: The Revolution in Modern Science.* (New York: Harper & Row Publishers, 1962).

p. 50: p. 49. / p. 55: p. 201. / p. 110: p. 186. / p. 114: p. 82. / p. 116: p. 71. / p. 123: p. 107. / p. 127: p. 114. / p. 131: p. 198-199.

Mehra, Jagdish, ed. *The Physicist's Conception of Nature.* (Boston: D. Reidel Publishing Company, 1973).

p. 66: p. 24.

Jeans, Sir James

Jeans, Sir James. *The Mysterious Universe.* (New York: MacMillan, 1932).

p. 79: p. 168. / p. 84: p. 181-182.

Jeans, Sir James. *Physics and Philosophy.* (New York: Dover, 1981).

p. 95: p. 193-194. / p. 98: p. 203. / p. 99: p. 193. / p. 109: p. 216.

Lamrimpa, Gen

Lamrimpa, Gen. *Realizing Emptiness.* Wallace, B. Alan, tr. (Ithaca, NY: Snow Lion, 1999).

p. 84: p. 85. / p. 85: p. 66. / p. 112: p. 68.

Minkowski, H.
Einstein, Albert. *The Principle of Relativity.* (New York: Dover, 1952).
p. 133: p. 75.

Murti, T. R. V.
Murti, T. R. V. *The Central Philosophy of Buddhism.* (London: Allen & Unwin, 1955).
p. 126: p. 198.

Nagao, Gadjin M.
Nagao, Gadjin M. *Madhyamika and Yogacara.* (Albany, NY: SUNY Press, 1991).
p. 144: p. 201.

Nagarjuna, Siddha
Murti, T. R. V. *The Central Philosophy of Buddhism.* (London: Allen & Unwin, 1955).
p. 122: p. 138.
Nagarjuna, Siddha. *Elegant Sayings.* (Berkeley, CA: Dharma Publishing, 1977).
p. 34: p. 38.
Nagarjuna, Siddha. *The Fundamental Wisdom of the Middle Way.* Garfield, Jay L., tr. (New York: Oxford University Press, 1995).
p. 46: p. 40. / p. 128: p. 14-15.
Ramanan, K. Venkata. *Nagarjuna's Philosophy as Presented in the Maha-Prajnaparamita-Sastra.* (Delhi: Motilal Banarsidass, 1975).
p. 132: p. 83.

Oppenheimer, J. Robert
Oppenheimer, J. Robert. *Science and the Common Understanding.* (New York: Simon & Schuster, 1954).
p. 44: p. 69. / p. 159: p. 8-9.

Pabongka Rinpoche
Tsongkapa. *The Principle Teachings of Buddhism.* Tharchin, Geshe Lobsang, tr. (Howell, NJ: Mahayana Sutra and Tantra Press, 1988).
p. 82: p. 119. / p. 114: p. 125.

Pauli, Wolfgang
Heisenberg, Werner. *Physics and Beyond.* (New York: Harper & Row, 1971).
   p. 59: p. 209.
Pauli, Wolfgang. *Writings on Physics and Philosophy.* (Berlin: Springer-Verlag, 1994).
   p. 48: p. 259. / p. 71: p. 41-42. / p. 163: p. 260.

Planck, Max
Planck, Max. *Where Is Science Going?* (Woodbridge, CT: Ox Bow Press, 1981).
   p. 35: p. 67. / p. 57: p. 106. / p. 82: p. 92. / p. 152: p. 198. / p. 161: p. 214.

Rabjampa, Longchen (Longchenpa)
Rabjampa, Longchen (Longchenpa). *The Four-Themed Precious Garland.* (Dharamsala: Library of Tibetan Works and Archives, 1979).
   p. 93: p. 45.

Rabten, Geshe
Rabten, Geshe. *Advice from a Spiritual Friend.* (London: Wisdom, 1984).
p. 31: p. 66.

Ramana Maharshi, Sri
Ramana Maharshi, Sri. *The Spiritual Teachings of Ramana Maharshi.* (Boulder, CO: Shambhala, 1972).
   p. 30: p. 72. / p. 79: p. 13.

Ramanan, K. Venkata
Ramanan, K. Venkata. *Nagarjuna's Philosophy as Presented in the Maha-Prajnaparamita-Sastra.* (Delhi: Motilal Banarsidass, 1975).
   p. 66: p. 90. / p. 127: p. 200.

Schrödinger, Erwin
Schrödinger, Erwin. *My View of the World.* (Woodbridge, CT: Ox Bow Press, 1983).
   p. 129: p. 22. / p. 154: p. 18.

Schrödinger, Erwin. *What Is Life & Mind and Matter.* (Cambridge: Cambridge University Press, 1967).

    p. 30: p. 154. / p. 64: p. 132. / p. 69: p. 137. / p. 72: p. 127. / p. 96: p. 156. / p. 155: p. 146.

Schrödinger, Erwin. *What Is Life? & Other Scientific Essays.* (Garden City, NY: Doubleday, 1956).

    p. 49: p. 165. / p. 164: p. 232.

Sengtsan

Sengtsan. *Hsin Hsin Ming: Verses on the Faith-Mind.* Clarke, Richard B., tr. (Buffalo, NY: White Pine Press, 1973).

    p. 68: p. 7. / p. 124: p. 9. / p. 130: p. 10. / p. 150: p. 6.

Shankara

Alston, A. J. *Samkara on the Creation: A Samkara Source-Book, Vol. II.* (London: Shanti Sadan, 1980).

    p. 77: p. 123. / p. 94: p. 208. / p. 101: p. 38-39. / p. 142: p. 156. / p. 149: p. 202. / p. 154: p. 208.

Shankara. *Shankara's Crest-Jewel of Discrimination* (*Viveka-Chudamani*). Isherwood, Christopher, tr. (Hollywood, CA: Vedanta Press, 1971).

    p. 29: p. 98. / p. 37: p. 112. / p. 57: p. 101. / p. 70: p. 100. / p. 78: p. 97. / p. 96: p. 71. / p. 99: p. 98.

Stapp, H. P.

Stapp, H. P. "Are Superluminal Connections Necessary?" (*Nuovo Cimento*, vol. 40B, 1977).

    p. 141: p. 191.

Suzuki, D. T.

Suzuki, B. L. *Mahayana Buddhism.* (London: Allen & Unwin, 1959).

    p. 133: p. 33.

Suzuki, D. T. *On Indian Mahayana Buddhism.* Conze, E., ed. (New York: Harper & Row, 1968).

    p. 32: p. 70. / p. 35: p. 237. / p. 41: p. 49. / p. 59: p. 239. / p. 61: p. 56. / p. 71: p. 45. / p. 95: p. 63. / p. 108: p. 47. / p. 113: p. 47-48. / p. 116: p. 198. / p. 120: p. 167. / p. 129: p. 149. / p. 139: p. 60-61. / p. 110: p. 270.

Suzuki, D. T. *Zen Buddhism.* (Garden City, NY: Doubleday, 1956).
  p. 48: p. 84. / p. 83: p. 270. / p. 81: p. 269.

Tayé, Kongtrul Lodrö
Tayé, Kongtrul Lodrö. *Myriad Worlds: Buddhist Cosmology in Abhidharma, Kalacakra, and Dzogchen.* (Ithaca, NY: Snow Lion, 1995).
  p. 89: p. 202. / p. 98: p. 198.

Tsogyel, Yeshe
Tsogyel, Yeshe. *Sky Dancer.* Dowman, Keith, tr. (New York: Penguin, 1989).
  p. 72: p. 159. / p. 107: p. 151.

Tsongkapa
Tsongkapa. *The Principle Teachings of Buddhism.* Tharchin, Geshe Lobsang, tr. (Howell, NJ: Mahayana Sutra and Tantra Press, 1988).
  p. 138: p. 118.

Tzu, Chuang
Tzu, Chuang. *Chuang Tzu: Basic Writings.* Watson, Burton, tr. (New York: Columbia University Press, 1964).
  p. 25: p. 26. / p. 28: p. 42. / p. 47: p. 102. / p. 49: p. 34-35. / p. 55: p. 46. / p. 56: p. 99. / p. 92: p. 45. / p. 117: p. 39. / p. 118: p. 44.

The Upanishads
The Upanishads. Mascaro, Juan, tr. (New York: Penguin, 1965).
  p. 58: p. 51. / p. 135: p. 113. / p. 148: p. 64. / p. 153: p. 62-63.

Utpaladeva
Lawrence, David Peter. *Rediscovering God with Transcendental Argument.* (Albany, NY: SUNY Press, 1999).
  p. 73: p. 110.

Vivekananda
Vivekananda. *Jnana Yoga.* (New York: Ramakrishna-Vivekananda Center, 1972).
  p. 140: p. 109.

Weyl, Hermann
Weyl, Hermann. *The Open World.* (Woodbridge, CT: Ox Bow Press, 1989).

 p. 29: p. 28-29. / p. 108: p. 28. / p. 160: p. 28-29.

Weyl, Hermann. *Philosophy of Mathematics and Natural Science.* (Princeton: Princeton University Press, 1949).

 p. 101: p. 171. / p. 142: p. 153-154.

Wheeler, John A.
Wheeler, John A., ed. *Quantum Theory and Measurement.* (Princeton: Princeton University Press, 1983).

 p. 132: p. 204. / p. 145: p. 205-206.

Wigner, Eugene P.
Wigner, Eugene P. *Symmetries & Reflections.* (Bloomington, IN: Indiana University Press, 1967).

 p. 77: p. 189.

THOMAS J. MCFARLANE is "an independent scholar of no mean stature" in the words of eminent professor of religion and philosophy Huston Smith. His writings on science and religion have appeared in journals and conference proceedings, and he has given talks on the subject for over a decade. A graduate of Stanford University in physics, he also holds advanced degrees in mathematics and in philosophy and religion. In addition to his scholarly work, he has been immersed in the study and practice of the world's spiritual traditions for over 12 years. He lives in Eugene, Oregon.

WES NISKER, who penned the introduction, is also the author of two bestselling books, *Buddha's Nature* (Bantam) and the underground classic, *Crazy Wisdom* (Ten Speed Press). He has practiced Buddhist meditation for 30 years with various teachers in Asia and the West, and is the founder and editor of the bi-annual Buddhist journal *Inquiring Mind*. He teaches retreats and workshops in Buddhist meditation and philosophy at venues internationally, and is affiliated with Spirit Rock Meditation Center in Woodacre, California.